For my wife Bríd and children David, Joanna and Daniel

Acknowledgements

The publishers are grateful to Professor David Fegan, MRIA, and Professor Frank Hegarty, MRIA, for reading and commenting on the text and to Pauric Dempsey for his work on the project.

The Industrial Development Authority partnered with the Royal Irish Academy in the production of the book, allowing it to be produced in full colour and at an accessible price.

Images were provided efficiently and expertly by Mark McGrath at the picture desk of the *Irish Times*.

In particular, the publishers would like to express their appreciation to Geraldine Kennedy, Editor of the *Irish Times*, for encouraging the publication.

Note on the choice of articles

These articles are pieces originally published in the *Irish Times,* chosen to give a snapshot of the work being carried out by scientists in Ireland today and by Irish scientists abroad. They are not intended to be comprehensive and we regret any omissions or oversights.

Contents

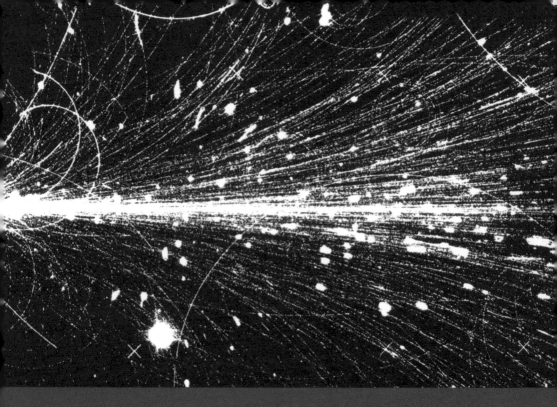

The cutting edge of Irish science

The research landscape in Ireland has been transformed over the past two decades. Twenty years ago state investment in research was almost non-existent. Now, enough money is available to encourage postgraduates to consider pursuing a research career within Ireland.

A little matter of €2,540 million arriving via the National Development Plan certainly helped matters along. In the mid-1980s budgets were being slashed in an arbitrary fashion. Not even medical research escaped cost cutting despite the potential benefits to patients. Now, world-class medical research is underway in purpose-built laboratories that link university to university but also university to hospital complex. The transfer of research findings from lab bench to bedside is certainly well underway here.

Admittedly, much of the change has come about only since 2000, when the €40-billion National Development Plan was announced. It ring-fenced unprecedented amounts of money for the conduct of scientific research in Ireland. It also opened the doors, both for the return of expatriate Irish scientists and for the arrival of scientists of any nationality interested in sharing in the government's research largesse.

The actual spark that brought about such rapid change, however, was the Technology Foresight initiative set in motion by government in March 1998.

It was surprising for a number of reasons, principally because of its independence but, more importantly, because it was told to cast ahead more than 15 years to decide what Ireland needed to do if it was to exploit the advance of future technology.

For those unfamiliar with the workings of government, this represented the lifetime of three full-term governments. The powers that be were looking for recommendations that would long outlive the lifetime of the government that had asked for them. The 2015 horizon represented one of the first clues that this time things were going to be different for scientific research in Ireland.

The Technology Foresight exercise recommended that Ireland focus its energies on two key areas: biotechnology and information and communications technology. It also suggested that support be given to the basic sciences that underpinned these areas. Most importantly, it called for the establishment of a Technology Foresight Fund that would help build up Ireland's research infrastructure.

Sure enough the fund was delivered when details of the National Development Plan 2000–2006 were announced. New funding bodies such as Science Foundation Ireland and the two research councils were created, and existing ones like the Higher Education Authority's Programme for Research in Third-Level Institutions and Enterprise Ireland were ramped up and given fresh supplies of money.

Scientists working in the third-level sector had been almost entirely dependent on EU sources for research funding. The people and the capacity to do good science were already available, however, and scientists here, to their credit, rose to the occasion. All they had been waiting for was the finance and a chance to pursue their scientific goals.

The result, after more than six years of National Development Plan funding, can be seen today in new, well-equipped labs, lots of research activity and an entirely new level of co-operative activity that links the various third-level institutions and their shared research goals. I started writing about Irish science during the dark days of the mid-1980s, when Irish scientists excelled at getting more financial support per capita out of the EU's various research programmes than anyone else in the Community. These are changed times, and now researchers here have the capacity to do world-class science.

The research community now also awaits sight of the government's next national plan and the science budget it might contain. Everyone is now convinced that the funding will continue and that scientific endeavour can go from strength to strength here. It makes the goal of a truly knowledge-based economic future that much more achievable.

Flashes of brilliance that change it all

The splitting of the atom, the description of DNA and the development of computers may have been the great scientific advances of our age, but their impact on society has depended on social, political and economic factors. Our failure to grasp this fact has led to a lack of understanding, and in some cases, our fear of science in this century.

No other human endeavour has a greater influence on society, a greater ability to bring about change, than science. Waging war comes a close second with its ability to reconstruct a population or redraw the political map, but war is almost exclusively a negative and wasteful undertaking, nothing like the creativity and innovation that flows from science.

Science is also an evolutionary process, punctuated by a few but remarkable revolutionary interjections, discoveries that don't as much nudge as kick the process forward. Every century has its scientific punctuation, the past 100 years perhaps more so than previous ones. In the past, discoveries had the potential to affect science as a study—today, they have the power to alter the way we live, the way our society is structured. However, any examination of the past century's science must start with some appreciation of what science actually is. Few people readily make

the connection between science and creativity. Many believe art and creativity lurk on one side of the brain while cold logic, mathematics and the pure sciences occupy the other. The popular perception is of two mutually exclusive halves, fighting for control to determine whether a person will write odes or don a white coat.

In fact, there is no effort more creative than research. Although science imposes a tightly controlled regime that dictates how experimentation proceeds and what conclusions can be drawn from it, the ideas that initiate the research spring from the same well of inventiveness and original thought that refreshes the poet, the novelist or the musician.

Many may perceive the self-imposed scientific tyranny that rules the laboratory and the research effort as a burdensome yoke that forces the scientist to conform to particular methods, no matter how bright the initial flash of a new idea. In reality, these strictures are no different from the despotism under which accomplished musicians must labour to achieve mastery of an instrument or total control over the magic of a wonderful voice. The novelist's accomplishment is also seen as inspiration. But to learn what it is really like to be a writer, take your absolute favourite novel, sit down and try to type it out. The mechanical labour that attaches to the process will show that delivering creativity in a way that demonstrates its true worth is hard work, whatever the discipline.

Genuine creativity, particularly when it touches that rare thing, genius, is recognisable as such no matter whether it is found in the arts or sciences. While you don't have to be an artist to see the beauty in Caravaggio's *The Taking of Christ*, or a musician to appreciate Mozart or Brahms, the self-same achievement is visible in Einstein's theory of relativity for those with eyes to see. His creative genius is less immediately visible to the casual observer. Yet Einstein's work and the creative efforts of other great scientists have had a more tangible and a more profound impact on society than any concerto or novel.

Great literature and art are wonders to behold, but they move only those with the inclination to appreciate them. No law has yet been passed to force people to exploit the treasures on display in the world's museums or to read the writings of Nobel laureates. As a result, these works are fully appreciated by relatively few people. Science, however, manages to turn society around, even if the individuals in society remain ignorant of or indifferent to its discoveries.

It doesn't matter if you understand Einstein's little formula $E = mc^2$, even though it tells you how big a bang to expect from a nuclear warhead. Know for certain, however, that humanity has lived under the spectre of mutually assured nuclear destruction for almost two generations and no other influence has had a greater impact on the geopolitics of our shared planet. Watson and Crick, meanwhile, might sound vaguely familiar, like the names of a comedy duo you once heard on the BBC, or the names of political commentators you should probably know. Yet their scientific discoveries about the nature of DNA, the genetic blueprint for life, sparked a revolution in biochemistry, which so far has delivered genetic engineering, genetically modified foods and a whole new way of looking at human health and disease.

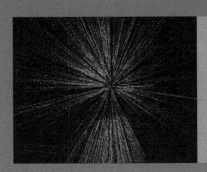

No other human endeavour has a greater influence on society, a greater ability to bring about change, than science.

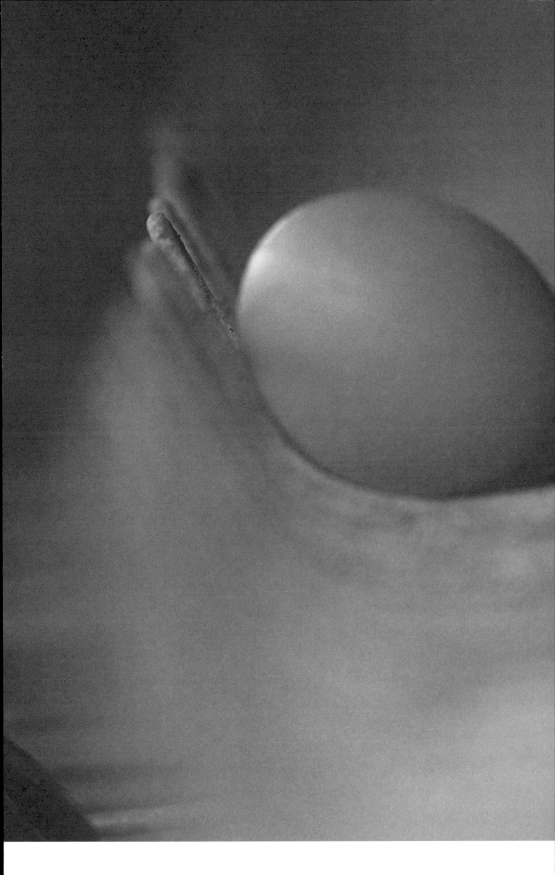

The chicken comes first

Researchers at St Vincent's Hospital Dublin finally have an answer for the old question about the chicken or the egg coming first.

If it comes down to a vote, the Education and Research Centre (ERC) at St Vincent's Hospital Dublin is very definitely on the side of the chicken. The egg is all very well but the chicken comes first. This favouritism arose during the course of an ongoing research project into unique anti-microbial substances known to exist in the yolk of an egg. These novel substances could have an important role in the control of infections and as a way to defeat antibiotic resistant bacterial strains, explains the Centre's research director, Dr Cliona O'Farrelly.

"What we are looking for are novel anti-microbial factors," she says. Of course, when you go looking for something you often find something completely different and this was the case with her team's work. They found a way to make some microbes come unstuck, preventing them from having a harmful effect.

Most animals, including humans, have an advanced immune system that deploys antibodies to defeat infections. Organisms such as lobsters, insects and other less complex species don't have antibodies but can defend themselves from bacterial infection. They rely on an 'innate immune response', an ancient form of immune defence that predates the acquired antibody response on the evolutionary chain. It involves the very rapid release of powerful anti-microbial substances, peptides that can recognise and attack invading bacteria. "They are hard-wired to recognise harmful pathologic structures," Dr O'Farrelley explains—for example, responding to substances on the outer bacterial coat.

It was known that eggs must have some form of anti-microbial factors too. The ERC had discovered that calves fed raw egg seemed to be protected against salmonella infection. Researchers elsewhere were also starting to find these "collective anti-microbial peptides" in other species, she said. "What was being discovered was small peptides with antibacterial activity."

The Centre received a Department of Agriculture and Food grant under the Food Institutional Research Measure programme and began the search for these substances in egg yolk. "We found anti-microbial activity in lipoproteins in eggs," she says and last month published these findings in the *Journal of Food Science*.

The substance's anti-microbial response was relatively weak against pathogens tested so far other than *Streptococcus mutans*, the organism involved in causing tooth decay. The team was puzzled by the selectivity of the peptides. "We wondered if there might be something in them that prevented bacterial adhesion."

The ERC joined in a collaborative research project with Mr Mike Folan, founder of Westgate Biological Ltd, to try and understand what might be interfering with bacterial adhesion. Westgate develops healthcare products derived from eggs and milk.

As part of this collaboration, the partners used a calcium carbonate model of *S mutans* infection of teeth. This showed that in fact the factors discovered in eggs did interfere with the bacteria's ability to stick to a surface. Adhesion is a key characteristic of many successful bacteria, without which they don't have an effect. "What we showed was that it prevented adhesion and later found that it could prevent salmonella adhesion to epithelial cells [cells found in many parts of the body including the intestines]," she explains.

Using an anti-adhesion factor to block bacteria "is all quite new as an idea", she says, and might offer a way to stop bacteria that are resistant to antibiotics. "The problem is if you give antibiotics, the bugs find ways of escaping by mutating." Anti-adhesion factors might prevent this escape because they don't provoke mutation.

Dr O'Farrelly went back to the drawing board—or in fact the chicken—when considering these findings. "What we found was even hens which had not been vaccinated against e coli had resistance." Eggs clearly contain anti-microbial factors, but so too must the chickens, probably arising in their innate rather than their antibody immune response.

"We suddenly realised we should have started with the chicken and not the egg," she says. She applied for—and got—fresh funding from the Department's programme and started a new study in cooperation with Dr Grace Mulcahy, dean of research, and Prof. Alan Baird, both of UCD's Veterinary College.

The search is now underway for the chicken genes responsible for producing these innate anti-microbial factors. Tracking them down is a challenge, admits Dr O'Farrelly. The team uses a back-to-front approach, looking not for the unknown protein but for the tell-tale genetic instructions, RNA, produced when a gene has been switched on.

The team searches for RNA 'expression sequence tags' and then analyses them using advanced bioinformatics computer systems. This will help the scientists to backtrack to the gene that produced the tag and finally to the anti-microbial factor it produces.

Thursday, 9 May 2002

➕ **St. Vincent's Hospital Education and Research Centre**: http://www.st-vincents.ie/erc_intro.html
UCD Faculty of Veterinary Medicine: http://www.ucd.ie/vetmed/
Westgate Biological Ltd: http://www.westgate.ie/

D. Brady, S. Gaines, L. Fenelon, J. Mc Partlin and C. **O'Farrelly,** (October 2002) A lipoprotein derived antimicrobial factor from hen egg yolk is active against *Streptococcus* species, *Journal of Food Science* 67(8), 3096–3103.

Journal of Food Science: http://bookstore.ift.org/store/iftstore/newstore.cgi?itemid=20327&view=item&categoryid=557&categoryparent=557&page=1&loginid=4783439

Reasons for standing tall

Research groups at Trinity College Dublin and Iowa State University have independently confirmed a 30-year-old theory about how animals with backbones may have evolved. Both groups point to a sudden unprecedented increase in genetic complexity, in turn leading to a new family of animal species.

Computing power and data from the human genome project have allowed Trinity College researchers to support an old theory about how animals with backbones may have arisen.

Susumu Ohno developed the theory in a 1970 book, *Evolution by Gene Duplication,* but his ideas lay idle and unproven. He suggested that the greater complexity of animals with backbones occurred because of a dramatic doubling of the number of genes in their genetic code through a process called polyploidy.

Polyploidy is a process where a genetic mistake, either during cell division or when male and female sex cells fuse, causes a doubling of the genome of a species. Instead of a creature having the same number of chromosomes and genes as its parents, the amount of genetic material doubles up, encouraging the emergence of completely new organisms. Polyploidy is nothing new in the plant kingdom and it is also known in fish, but was unknown in animals with spines—chordates. Polyploidy gave rise to a new fish species, the salmon, as recently as 10 million years ago. About 70 per cent of flowering plants and 95 per cent of ferns underwent polyploidy. Now the Trinity and Iowa groups demonstrate at least one major event of polyploidy in chordates, an occurrence that may have allowed the greater complexity of backboned animals to arise. Their separate research findings are published in the current issue of *Nature Genetics*.

Work at Trinity started three years ago with funding from the Health Research Board. Science Foundation Ireland funding then came on stream about a year ago. Ohno knew that polyploidy was at work in the fish DNA, explains Prof. Ken Wolfe, assistant professor of genetics at Trinity's Department of Genetics. "He could see these early polyploid events in the fish genome," he says and theorised something similar was at work in chordates. "It was an unprovable theory", however, and remained so until the human genome project laid bare the contents of the entire human genetic code. His lab was involved in the genome project and sequenced elements of our DNA. "We have now used the human DNA sequence to test [Ohno's] theory," says Prof. Wolfe. "We have come in and said yes it looks like he was right."

The confirmation comes as a result of bioinformatics, the business of using computers to analyse the billions of steps in human DNA. The computers can scan for duplicate fragments, "just pieces, maybe a few million base-pairs long that match others on different chromosomes", explains Prof. Wolfe. The assumption was that if polyploidy occurred, then it happened hundreds of millions of years ago, before the chordate group emerged. It could have arisen in a single organism or the merger of two to form one with double the number of genes, says Prof. Wolfe.

His team, which included Aoife McLysaght and Karsten Hokamp, set up the kinds of matches to search for. The computers churned through the mass of data for six hours before returning a positive verdict.

The team found matching fragments well separated on different chromosomes, the longest including 29 genes and 40 million base-pairs. The researchers found many duplicated DNA sequences. "We have about 100 of them and they occupy about half the genome."

They also assessed when this might have occurred and suggest what is likely to have been a once-off 'point event' happened between 350 million and 650 million years ago. To put this timing in perspective, the divergence that separated humans and fish on the evolutionary tree occurred about 420 million years ago and the human/mouse divide is estimated at 100 million years ago. "We can't really say what triggered it, just that it happened," says Prof. Wolfe. "One of the interesting questions is what are the functions of these genes? The big advantage of duplicating genes is one of the genes can sit there as a back-up and can change." If the size of the chordate genome doubled, all the existing genes would have had a working match. That would have allowed the extra gene to specialise over time, and become something different from the original without harming the organism. This would have added to the range and function of the proteins being produced by these genes.

"One of the next things we want to do is look at the mouse genome," says Wolfe. The team will try to confirm polyploidy in its genome and this may enable them to refine a timing for the genetic doubling event. They will also see whether they can confirm another Ohno theory, that polyploidy happened twice in chordate history.

Thursday, 30 May 2002

+ **TCD Genetics Department:** http://www.tcd.ie/Genetics/

A. McLysaght, K. Hokamp and **K.H. Wolfe**, (2002) Extensive genomic duplication during early chordate evolution, *Nature Genetics* 31(2), 200–04.

Nature Genetics: http://www.nature.com/ng/journal/v31/n2/abs/ng884.html

Fixing our DNA
for the next generation

An NUI Galway team is studying a key system used by the body to repair damaged DNA and help prevent cancers.

Cells go to great lengths to protect the integrity of our genetic code. They use powerful systems that respond instantly whenever damage occurs to DNA, carrying out running repairs and even triggering spontaneous cell death if the damage is too great.

These systems are known collectively as the 'DNA damage checkpoint pathway'. The pathway co-ordinates a range of responses inside the cell when DNA is damaged by sunlight, chemicals or ionising radiation. "This pathway is crucially important for the prevention of cancer," explains Prof. Noel Lowndes of NUI Galway. The university is researching how the pathway works, and Lowndes is building a team to learn its secrets.

Lowndes returned to the Republic of Ireland from the Imperial Cancer Research Fund in Britain, one of the top cancer research organisations in the world. He now heads the 'Genome Stability Cluster', part of the National Centre for Biomedical Engineering Science at NUI Galway. The cluster received funding worth €2.7-million from the university, from the Programme for Research in Third-Level Institutions run by the Higher Education Authority and from the Health Research Board. The cluster has five lead researchers, including Dr Ciaran Morrison, who recently received an additional €1.4-million investigator award from Science Foundation Ireland. "What we are hoping to build here in Galway is a world-class centre for this subject," says Lowndes.

Our genetic code is under constant threat and assault, much of it—but not all—brought on by our own actions. Ultraviolet radiation from the sun can cause breaks in our DNA. Free oxygen from the foods we eat causes damage, as do cosmic rays from space and from natural and man-made radiation sources.

DNA is shaped like a ladder that has been twisted into a helix. The cell's own repair systems can fix single breaks in this ladder, but the pathway intervenes when a more serious, double-strand break occurs. The pathway is a handful of very important proteins circulating in the nucleus, and ready to spring into action. "You can think of it as a panic response," says Lowndes. The proteins react immediately to the damage, triggering a cascade of events that can fix things in minutes or within a few hours.

"The biochemistry of the pathway is still largely unknown," he says. "My particular interest is at the very top of the pathway, how does it detect the damage?" Once set in motion, the pathway has profound effects inside the cell. It can halt cell division for up to a day or two while repairs occur. It can increase the effectiveness of the repair systems by changing or increasing the output of repair proteins. And if a cell cannot be repaired, it triggers controlled cell death, apoptosis.

that cells which go wrong are either corrected or killed off.

The Galway team is looking, however, at what happens when the initial instability occurs in genes responsible for the protective pathway proteins. "Mutation of pathways that control genome stability has been shown to be an absolute prerequisite to cancer," says Lowndes.

"Once you set up just the right amount of genomic instability, within a few decades you can develop the six to seven mutations to create a cancer. If you mutate one of the

+ Professor Noel Lowndes

Interest in the pathway is strong because its failure can result in the development of cancers. Unrepaired damage can cause mutations in the DNA, errors that are carried into subsequent generations of a cell. The initiating 'genome instability' can cause later mutations, and it is believed that at least six or seven such mutations are needed to convert a normal cell into a cancer cell, he says. Normal cells divide with such accuracy that it would be virtually impossible for a single cell to accumulate all the necessary mutations needed to become cancerous. The pathway proteins ensure

very guardians of genomic stability then you set up genomic instability." Studying the pathway in detail represents a considerable challenge, however. Its key initiating proteins are not abundant, so getting enough protein for biochemical analysis is difficult, says Lowndes.
Understanding exactly how it works could lead, however, to important new drugs that could boost or block elements of the pathway as a way to prevent or treat cancers.

Thursday, 6 June 2002

 NUI Galway National Centre for Biomedical Engineering Science: http://www.nuigalway.ie/ncbes
Imperial Cancer Research Fund: http://www.icnet.uk/

Zebrafish may reduce the risks of heart attacks

An Irish researcher in the US has come up with an unusual new ally in the fight against human heart disease— the zebrafish.

What do the zebrafish and human heart disease have in common? Quite a lot, according to a research team at the University of Pennsylvania in Philadelphia. The scientists, headed by an Irishman, found that the Zebrafish produces two proteins very similar to human proteins that have important connections to heart disease. This makes the zebrafish important to future studies of drugs that can reduce the risk of heart attack in humans.

Prof. Garret FitzGerald, originally from Dublin, is chairman of the department of pharmacology at Penn and director of the Penn Centre for Experimental Therapeutics. He, lead author Dr Tilo Grosser and colleagues published their findings about the zebrafish this month in the *Proceedings of the National Academy of Sciences*. These findings are the result of many years of work by FitzGerald with two key human enzymes, cyclooxygenases, known as COX-1 and COX-2. They play a role in a variety of ailments, including cardiovascular disease, some types of cancer and arthritis.

The proteins are like the two ends of a seesaw, performing a balancing act that increases or decreases the stickiness of blood and the diameter of the blood vessels. They also play a role in inflammation. COX-1 is found in platelets, the blood cells that form clots. It produces a substance that makes platelets sticky and causes blood vessels to constrict. These are useful

+ Translucent Zebrafish embryos

responses to help close a cut, but become life-threatening as a prelude to heart attack or stroke. COX-2 is expressed in blood vessels and produces a substance that opens up the vessels and blocks the activation of platelets.

These two enzymes are the targets of various drug therapies. Aspirin blocks the action of COX-1, can thin blood and is a painkiller useful in inflammation. It can also, unfortunately, cause stomach irritation that can lead to ulcers. Drugs that block COX-2 can decrease the pain associated with inflammation and are therefore often used to control the pain of arthritis. They do so without causing stomach irritation, which is an advantage, but it remains unclear whether tampering with the balance between the two COX enzymes could cause problems.

Last April, FitzGerald, lead author Dr Yan Cheng and colleagues at Penn published an explanation of the interaction between COX-1 and COX-2 in the journal, *Science*. They wanted to investigate "the interplay of the two COX products in the cardiovascular system", FitzGerald explains. They found that the COX-2 product, prostacyclin, played a vital role in restraining the harmful cardiovascular effects of the COX-1 product, thromboxane. The team used mice bred so that their systems couldn't use prostacyclin, imitating the effect of a COX-2 inhibitor. The researchers discovered that the mice had a greatly exaggerated response to injury and activation of their platelets, leaving them at greater risk of unwanted blood clots.

Enter the zebrafish. The fish produce enzymes that match the two COX enzymes, and the new research shows that they respond to drugs in a way similar to the human enzymes. "We have learned a great deal about how the COX enzymes and their inhibitors work from mouse models of COX gene inactivation. However, these systems have their limitations," FitzGerald says. "The zebrafish promises to play a complementary role in which both biology and the role of drugs can be investigated."

The fish are likely to have a key part in the identification of drugs useful for the regulation of the COX enzymes, the researchers believe. "The zebrafish has particular advantages for the study of drug action," says the lead author in the zebrafish paper, Dr Grosser. "Their embryos are translucent, so we can study the pattern of gene expression during development, as well as in the adult. The near-completion of the zebrafish genome project allows us to hunt for relatives of human genes of interest. Then we can manipulate them and see how they function."

While any new drug will still have to go through detailed safety tests, at least researchers will have good solid clues about their likely effects and whether they are suitable to the job. This could mean safer, more effective pain control and a reduction in heart attack risk for patients.

Thursday, 27 June 2002

✛ **Center for Experimental Therapeutics** (now part of the Institute for Translational Medicine and Therapeutics): http://www.itmat.upenn.edu/index.shtml

Tilo Grosser, Shamila Yusuff, Ellina Cheskis, Michael A. Pack and Garret A. **FitzGerald,** (June 2002) Developmental expression of functional cyclooxygenases in zebrafish, *Proceedings of the National Academy of Sciences* 99, 8418–23

Proceedings of the National Academy of Sciences: http://www.pnas.org/cgi/content/abstract/99/12/8418

Yan Cheng, Sandra C. Austin, Bianca Rocca, Beverly H. Koller, Thomas M. Coffman, Tilo Grosser, John A. Lawson and Garret A. **FitzGerald,** (April 2002) Role of prostacyclin in the cardiovascular response of Thromboxane A_2, *Science* 296, 539–41.

Science: http://www.sciencemag.org/cgi/content/abstract/296/5567/539

How an Irish biochemist became the most cited computer scientist

Dr Des Higgins's work merges two disciplines at what has become a very active area of research—bioinformatics.

An Irish biochemist holds the unusual distinction of being the most cited computer scientist in the world. Although he isn't even a computer scientist. This is one of a number of 'citation' records held by Dr Des Higgins, a lecturer in the Department of Biochemistry at University College Cork.

A citation occurs when a scientist publishing research work in a journal mentions earlier work by another scientist in the list of references. A US company, the Institute for Scientific Information, catalogues citation 'scores' and publishes them in a citation index. It lists Dr Higgins as the most cited computer scientist and the seventh most cited molecular biologist.

He is co-author on a 1994 paper with J.D. Thompson and T.J. Gibson, ranked as the number one paper for citations in any discipline published over the past 10 years. Two papers published while at Trinity College and one at UCC together have collected 6,000 citations for Dr Higgins, and his current running total for citations stands at about 14,500.

"It is quite bizarre," Dr Higgins says of the citation game. "It is a measure of the impact of your work, but it is a crude and poor measure. People can overrate citations, but people like it because it is an easy thing to measure."

His enormous catch of references occurs because his work merges two disciplines at what has become a very active research interface—bioinformatics. This refers to the use of computer power to help decipher the complexities of the human genetic blueprint, DNA. The 1994 paper described a computer programme called CLUSTAL W. "That particular paper is about a technique that biologists use to check DNA sequences. It was explaining a program we wrote," he says.

"I am a biologist by training but have been using computers since 1980. My PhD was in zoology in Trinity. For that I had to write a lot of computer programs in numerical taxonomy. Even though I was a biologist, I was working with computers. This has now hit the headlines in a big way through bioinformatics." He needed a computer program that could make comparisons between DNA sequences, cataloguing similarities and differences. CLUSTAL was

the result and the program now ranks as the second most commonly used software in DNA analysis.

When CLUSTAL was first developed the research team at Trinity's Department of Genetics gave it away free. "When we did that there was no money to be made," says Dr Higgins. This in turn made it extremely valuable and attractive to other researchers working in DNA analysis so CLUSTAL became a standard for this research.

BLAST is the most often used bioinformatics software. It takes a single given strand of DNA, enough to represent a single gene, and searches for matches anywhere else in the rest of an organism's genome. The DNA of other species can also be searched for matches, a common occurrence because nature has conserved many duplicate genes across species.

The human genome project showed that there are about 40,000 genes in our DNA each producing a unique protein. Yet "every gene in the human genome has many relatives", says Dr Higgins. CLUSTAL helps find these relatives and can spot differences between them.

His ongoing work involves further development of the multiple alignment method, finding ways to do more simultaneous alignments with longer DNA or amino acid strings. "We have to develop methods that are faster and more powerful," he says.

Thursday, 4 July 2002

➕ **UCD Conway Institute:** http://www.ucd.ie/conway

TCD Genetics Department: http://www.tcd.ie/Genetics/

Institute for Scientific Information: http://www.isinet.com/

J.D. Thompson, D.G. **Higgins** and T.J Gibson, (1994) CLUSTAL W: Improving the sensitivity of progressive multiple sequence alignment through sequence-weighting, position-specific gap penalties and weight matrix choice, *Nucleic Acids Research* 22(22), 4673–80.

Nucleic Acids Research: http://nar.oxfordjournals.org/cgi/content/abstract/22/22/4673

A true measure of pedigree

A Trinity College research group has discovered 'errors' in the bloodlines of international thoroughbred horses. The work points towards mistakes in the recording over centuries of *The Stud Book,* which details the family trees of pedigree horses.

Knowing the bloodlines of an animal is central to its designation as a thoroughbred, says Dr Emmeline Hill, a former postdoctoral research fellow in the Department of Genetics at TCD and lead author in a study of thoroughbreds published this month in the journal, *Animal Genetics.*

"People will pay millions of dollars for a good pedigree," she says. The current record price for a foal, based entirely on the animal's bloodlines, is $13.1 million, paid in the mid-1980s for an unbroken yearling yet to feel the burden of a saddle.

The Trinity research group, which included other teams in Egypt, Turkey, the US and Russia, was therefore surprised to find that the pedigree lines were somewhat clouded by mistakes. Some errors dated back to the establishment of the thoroughbred industry 300 years ago and others to the mid-1800s. The most recent, however, go back only 20 years, explains Hill.

The group looked at 100 animals in 19 'families' of horses. Each family was grouped along female lines, linking daughters to mothers to grandmothers and so on, back into the horse's family tree. "What we found in eight of the 19 families was there was at least one individual that didn't match the rest. That means that something happened in the history of the recording of the pedigree," she says.

Thoroughbred horse history is mapped out in *The Stud Book,* started by James Weatherby more than 200 years ago and still controlled by the Weatherby family. "The Stud Book was established in 1791. It is the bible of thoroughbred pedigrees. If your horse is not registered in Weatherby's, it's not a thoroughbred."

In the late 1800s an attempt was made to trace all mares then registered in the book back along their family trees to the foundation mares. These were grouped into families, ranked in order by the number of classic race winners the mare's bloodline had produced. Hill and colleagues looked at 19 of these families.

"The problem with *The Stud Book* and the recording of *The Stud Book* was that horses, particularly mares, weren't recorded particularly well," Hill said. A horse known by a vendor as the Old Grey Mare might on sale be quickly renamed. Mares were also often named after their sire and over the lifetime of a stallion there would be many of these.

The team decided to follow the female lines by studying mitochondrial DNA, genetic material passed from mother to daughter. Within each family, all females should have the same mitochondrial DNA. "Of the 19 families we looked at, about half of them had errors. Three or four of them are deep-rooted to the early days of *The Stud Book* when it would have been expected that errors would occur."

For example, family nine is linked to the Old Spot Mare born in 1700. There were 10 modern horses in the sample attributed by *The Stud Book* to Old Spot Mare, but in fact the majority, six, had the same mitochondrial DNA as horses in family 12, attributed to D'Arcy's Chestnut Arab Mare, also born in 1700. These findings will provide valuable insights into the earliest days of the throughbred, Hill believes. The work should also be of interest to the modern industry. "This could have consequences for thoroughbred breeders and buyers who often make million-dollar decisions in the sales ring based on the integrity of these pedigrees," she adds.

"I am not surprised by their findings in view of the fact that there are accidental mix-ups in the current foal crop," stated Weatherby Ireland's MD, Joe Kiernan. If there are mix-ups today they were there in the past. Recording errors are impossible since 1986, however, because all foals are DNA tested for parentage.

15 August 2002

⊞ **UCD Agriculture and Food Science Department:** http://www.ucd.ie/agri/
TCD Smurfit Institute of Genetics: http://www.tcd.ie/Genetics/

E.W. **Hill,** D.G. Bradley, M. Al-Barody, O. Ertugrul, R.K. Splan, I. Zakharov and E.P. Cunningham, (August 2002) History and integrity of thoroughbred dam lines revealed in equine mtDNA variation, *Animal Genetics* 33(4), 287.

Animal Genetics: http://www.blackwell-synergy.com/doi/abs/10.1046/j.1365-2052.2002.00870.x

The Stud Book: http://www.weatherbys-group.com/services/studbook.html

+ Dr Emmeline Hill

New treatments sought for muscular dystrophy

NUI Maynooth has begun a new research project looking at changes in the cell caused by the crippling disorder, muscular dystrophy.

An Irish research team hopes to find new treatments for Duchenne muscular dystrophy by studying how the disease damages muscle cells. Understanding the chemical changes caused by the disease should lead to new drug therapies for this degenerative condition.

Muscular Dystrophy Ireland has announced funding for a three-year research project to be headed by Dr Kay Ohlendieck, who takes over this autumn as professor of biology at NUI Maynooth. The project will build on work completed by Ohlendieck while working in University College Dublin's Conway Institute of Biomolecular and Biomedical Research, and published earlier this year in the *Journal of Applied Physiology*.

Muscular dystrophy is a collective term for a variety of neuromuscular conditions, and Duchenne is but one form. All forms are hereditary and characterised by the progressive degeneration and weakening of muscles. Duchenne is the most frequent human gender-specific inherited disease in the Republic of Ireland, occurring in one in 3,000 live male births.

Ohlendieck's UCD research showed what was happening inside muscle cells in Duchenne patients. The work was funded by the EU, the Health Research Board and Enterprise Ireland. The cells lose their ability to handle calcium properly. Calcium is essential for the messaging system used by muscle cells, says Ohlendieck. "If anything goes wrong with this calcium sequencing, the muscle loses strength." Although the primary genetic abnormality in this disease was previously established, little was known about what was happening in the cells and whether this might offer new treatments. The team at UCD included Kevin Culligan, Louise Glover, Paul Dowling and Niamh Banville. Ohlendieck's group established that the Duchenne patients' muscle cells were unable to control calcium.

The primary cellular change involves a weakening of the cell surface, says Ohlendieck. For a time the cells seem to be able to cope, but symptoms of muscle weakening

+ Professor Kay Ohlendieck, centre of photograph, with his team at NUI Maynooth, Pamela Donoghue, PhD student, left, Dr Daniela Schreiber and Phillip Doran, PhD student

begin to emerge when the patient is from three to five years old. "At some stage the regenerative ability of the cell doesn't work any more," he says. The weakened cell membrane lets calcium ions leak into the cell, raising levels and disturbing signalling. "Normally, the calcium levels inside the cell are very low. We thought maybe this failure in calcium regulation caused other effects." His group found that these cells couldn't clear the excess calcium, a job usually done by calcium regulating proteins. The calcium is usually moved into a storage area inside the cell, but in Duchenne patients this too is altered. "Besides the leaking of the surface membranes, the calcium stores have a lower capacity. More calcium fluxes in and the cell can't remove it to its internal stores."

The team developed a library of antibodies, an element of the immune system designed to target and help identify the calcium regulating proteins. They found three of importance, CLP 150, CLP 170 and CLP 220. They examined changes in the function of these proteins in those with and without the disease and studied the protein levels.

This work will now continue at Maynooth, with funding from Muscular Dystrophy Ireland, says Ohlendieck. He will use proteomics, the study of all the proteins produced in muscle cells, to examine abundances, interactions and so forth. "That might lead to the discovery of a new therapeutic target," he says. The funding was granted, in particular, for "screening for new therapeutic targets in calcium ion handling", he adds. The object would be to correct abnormal calcium cycling to reduce or eliminate dystrophic symptoms. There might be an opportunity to boost the calcium removal system or repair the surface membrane and so keep excess calcium out, says Ohlendieck. These options could emerge with a better understanding of the complex chemistry taking place inside the cells.

Thursday, 29 August 2002

➕ **NUI Maynooth Department of Biology**: http://www.nuim.ie/academic/

Conway Institute of Biomolecular and Biomedical Research: http://www.ucd.ie/conway/

Kevin Culligan, Niamh Banville, Paul Dowling and Kay **Ohlendieck**, (February 2002) Drastic reduction of calsequestrin-like proteins and impaired calcium binding in dystrophic mdx muscle, *Journal of Applied Physiology* 92, 435–45.

Journal of Applied Physiology: http://jap.physiology.org/cgi/content/abstract/92/2/435

Why nothing is the matter

An Irish physicist played a key role in the creation of 'large' amounts of antimatter at the CERN accelerator on the French/Swiss border. It marks a major breakthrough in particle physics.

A research group on the continent has racked up a first in the world of particle physics, the creation of 'large' amounts of antimatter. The team has so far produced about 50,000 atoms of 'antihydrogen', the mirror image of ordinary hydrogen.

The ability to produce so much antihydrogen is unprecedented, but it is still a tiny amount of antimatter, explains Dr Paul Bowe, a technical co-ordinator of the ATHENA antihydrogen project at CERN, the European Organisation for Nuclear Research. At that rate of production it would take the team a billion years to accumulate just one gram, he says. Yet it is a genuine breakthrough in the study of antimatter, that rare substance that for inexplicable reasons is very difficult to find in our universe. "It really is a first to do it," Bowe says. "It is an opportunity, it is a new window on the world of antimatter."

Bowe is part of an international team of physicists working at the Antiproton Decelerator facility at CERN. The team developed a technique and equipment to produce antihydrogen, the ATHENA collaboration, which stands for AnTiHydrogEN Apparatus. It involved 39 scientists in nine institutions and first started producing antihydrogen last August. Details of the research were published last month in the journal, *Nature*.

Antimatter smacks of Star Trek and science fiction, but the research is deadly serious. Cosmologists who try to explain the formation of the universe see no reason why our world should be made of matter as opposed to antimatter. "We know in the standard model of the universe that in the Big Bang equal amounts of matter and antimatter were produced," explains Bowe. Material condensed to form stars after the Big Bang but for some reason the antimatter disappeared. "We evolved as a matter world, there doesn't seem to be antimatter about," says Bowe. "Why? Where has it gone?" Having a collection of antimatter atoms on hand should help to answer these questions, he believes. "It becomes a little laboratory" where very big physics—the universe itself—can be studied.

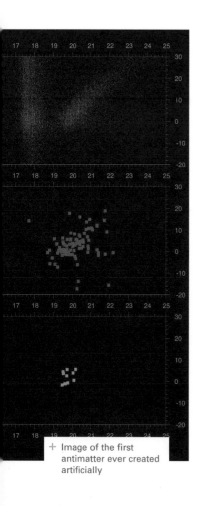

+ Image of the first antimatter ever created artificially

To understand antihydrogen it is necessary to reprise the structure of ordinary hydrogen, the most abundant element in the universe. Hydrogen atoms are the simplest of atoms, made up of a proton and an electron. Protons carry a positive charge while electrons carry a negative charge. Antihydrogen is a mirror of this, says Bowe. "Antihydrogen is a very simple thing, antielectrons, which are called positrons, and antiprotons." It carries a mirror image charge, with a negative antiproton and a positive positron, as its name suggests.

As explained in all the best sci-fi movies, put matter and antimatter together in one place and both are immediately destroyed with the release of large amounts of energy. This powerful reaction makes the creation of antimatter a challenge, but one that has been overcome by the CERN team.

Positrons are simple to produce, they are a natural radioactive decay product and large numbers of them are collected from a Sodium 22 source and trapped in a magnetic field. Antiprotons are more of a challenge, says Bowe. The CERN accelerator is used to smash high-speed protons into copper or iridium and antiprotons are produced from these collisions. They are moving very quickly, however, and must be slowed from close to the speed of light to a more manageable speed, about 400 to 500 metres per second. The Antiproton Decelerator at CERN can spit out about 10,000 positrons every two minutes or so.

The team produces about 70 million positrons and then allows the slowed, 'cooled' antiprotons to flow through the positron cloud, says Bowe. "It is then a matter of waiting and statistics."

Electromagnetic fields can trap the charged positrons and antiprotons, but when antihydrogen is formed it has no charge and drifts away. It is immediately annihilated when it touches matter, but this reaction is spotted by another device developed by the team, the "antihydrogen annihilation detector", says Bowe. The research results showed that several atoms of antihydrogen are formed every second using this process.

Antimatter was made previously at CERN and at the Fermilab accelerator in the US, but these atoms disintegrated immediately in the accelerator. The new CERN method will allow closer study of the atoms, Bowe explains. It could take another four years before the researchers develop useful methods for trapping the antimatter for study he says.

The antihydrogen will allow physicists to test the bedrock foundations of modern physics. One of the first things people will want to know is whether antimatter falls 'up' in the presence of gravity. And does time run backward in an antimatter world or forward. These things should apply across the board and having antimatter available will allow this to be proven. The charge-parity-time symmetry theorem is fundamental to our understanding of particles and how they interact as described by the standard model, Bowe says. If this symmetry theorem were found to be violated by antihydrogen the standard model would have to be revised. "Any deviation from this would be enormous, we would have to go back and start the standard model all over."

Thursday, 10 October 2002

➕ **CERN:** http://public.web.cern.ch/Public/Welcome.html

M. Amoretti, C. Amsler, G. Bonomi, A. Bouchta, P. **Bowe**, C. Carraro, C.L. Cesar, M. Charlton, M.J.T. Collier, M. Doser, V. Filippini, K.S. Fine, A. Fontana, M.C. Fujiwara, R. Funakoshi, P. Genova, J.S. Hangst, R.S. Hayano, M.H. Holzscheiter, L.V. Jørgensen, V. Lagomarsino, R. Landua, D. Lindelöf, E. Lodi Rizzini, M. Macrì, N. Madsen, G. Manuzio, M. Marchesotti, P. Montagna, H. Pruys, C. Regenfus, P. Riedler, J. Rochet, A. Rotondi, G. Rouleau, G. Testera, A. Variola, T.L. Watson and D.P. van der Werf, (October 2002) Production and detection of cold antihydrogen atoms, *Nature* 419, 456–59.

Nature: http://www.nature.com/nature/journal/v419/n6906/abs/nature01096.html

+ Dr Hilary McMahon

Stress test for CJD

Stress inside the cell may be a key factor in the development of BSE in cattle and the human form of the disease, Creutzfeld Jacob Disease (CJD). A Dublin-based research group hopes to use this finding to develop new ways to treat these diseases.

A UCD research group is looking for new treatments for animal and human forms of mad cow disease.

The tragedy of BSE was a frightening reminder about how suspect food production practices could lead to illness and death. The disease arose when animal waste left over after factory processing was in turn reconstituted into animal feed and given to livestock. Infected cattle developed the invariably fatal disease, but in turn that disease jumped across to humans, creating a wholly new form of CJD.

There have been two deaths from this new form of CJD in Ireland to date, but almost 150 in Britain. The eventual number of cases we are likely to see remains unknown, but studies in the UK suggest it could range from more than 10,000 up to 250,000 for the most pessimistic estimates.

The research by University College Dublin's 'prion group' within the Department of Industrial Microbiology therefore becomes very important. The team, led by Dr Hilary McMahon, hopes to find ways not just to halt the development of the disease in animals and humans, but to reverse it. Dr McMahon returned from a research centre in France to take up a post with the prion group in her alma mater, UCD. The Health Research Board, the Irish Research Council and Enterprise Ireland fund the work.

"We are trying to identify what factors are involved in the disease and how to cure them," McMahon says. The team has already enjoyed some success, pinpointing an unexpected cause of BSE. "One of the factors involved in the disease is oxidative stress." The cattle and human forms of the disease and the form found in sheep—scrapie—are known collectively as transmissible spongiform encephalopathies. The disease attacks brain cells and causes sponge-like holes that give it its name. All three forms of the disease are also linked to a change in the 'prion' protein that occurs naturally in brain cells. The disease converts the normal short-lived prion protein into a non-degradable form that builds up as plaques in the cells. It is known that the normal prions convert to the abnormal form when the two come in contact, thus causing a spiral of disease.

Reactive chemicals involving forms of oxygen that attach to and damage structures inside the cell cause the stress being examined by McMahon's group. These oxygen radicals arise naturally when we digest foods but are also produced by exposure to radiation and chemicals and by other causes.

The body protects itself by producing enzymes that mop up the 'superoxide' radicals explains McMahon. "A key enzyme is superoxide dismutase. This is one of the most important enzymes to remove superoxide from the cell." She is using *in vitro* methods to examine how stress levels in the cell are associated with the progression of scrapie. "Scrapie is a very good model for looking at other spongiform encephalopathies."

Brain cell lines are used to study the biochemistry behind the disease, tracing each step as cells are overcome. "We found that oxidative stress was involved in the process", says McMahon. Infected cells had increased levels of 'stress molecules', the superoxide forms that can damage proteins and substances inside the cell. At the same time levels of superoxide dismutase fell by half in the tests done so far. So, superoxide levels rose even as levels of the key protection enzyme fell once the disease process was introduced. McMahon and other groups have now found that the stress molecules alone can cause normal prions to convert into the diseased forms associated with scrapie. There is now speculation, she said, that stress molecules alone might actually cause the rare one or two spontaneous cases of CJD that arise in the Republic of Ireland in a typical year.

She hopes to find ways to interfere with this conversion process as a way to stop and then reverse the disease. "We are focusing on modifying the trafficking of the normal prion protein," she says. McMahon is also looking at the introduction of 'scavengers' to mop up excess oxidative molecules in the cell. "If you can identify a drug that can reverse it you would hope it could also prevent the disease".

Thursday, 28 November 2002

UCD Department of Industrial Microbiology:
http://www.ucd.ie/indmicro/html/overview.html

Scanning food for freshness

A team of Irish scientists has come up with the ultimate in food packaging—it tells you whether the food is fresh. A simple hand-held scanner lets you know whether the packaging is intact and the food is safe.

Researchers at Dublin City University have developed a new system of food packaging, which can tell you whether the food inside is fresh.

The Dublin City University research group believes the test system, Intellipak, could be worth billions of euro to food processors worldwide. It allows on-the-spot testing—either on the production line or in the home—without having to open the pack or destroy the product. The key to the DCU approach is a barcode-like test strip that provides the information which is built into the packaging itself, explains the head of the research group behind Intellipak, Prof. Brian MacCraith.

Consumer demand for fresh convenience foods has led food producers to develop new kinds of packaging that can protect fresh food from bacterial contamination, says MacCraith. "Increasingly, a lot of consumer food is packed in packages with modified atmospheres," he says of packs where the oxygen has been removed and replaced with gases such as nitrogen and carbon dioxide. "The idea is to extend shelf life," he adds. "Carbon dioxide actually retards spoilage organisms."

However, this modified atmosphere packaging (MAP) must remain intact, right through production, shipping, storage on retail shelves and all the way home to the

consumer. Even a tiny leak would let in oxygen, which aids the growth of the bacteria that cause spoilage. The challenge, however, is how can you tell if the packaging is intact?

The current practice, MacCraith said, involves random destructive testing. A pack is taken off the production line and its oxygen/carbon dioxide gas mix is measured. If the sample is bad it may mean having to destroy the complete batch or repackaging the entire run, a wasteful loss that can run into hundreds of thousands of euro, says MacCraith. There is also the risk that packaging failures will not be picked up in random testing, only to reach the marketplace with a now meaningless 'sell by' date. If the MAP has failed, then its sell by date will not be correct and the food will spoil more quickly.

"Our solution was getting rid of the random test and the destructive test by using a sensor film inside the pack," states MacCraith. The Intellipak was developed at DCU's National Centre for Sensor Research (NCSR). MacCraith is the director of this multi-disciplinary centre, which involves 25 research academics and 130 full-time researchers, including about 100 post-graduate and post-doctoral researchers.

+ Professor Brian MacCraith

The €12-million centre was funded by a grant from the government's Programme for Research in Third-Level Institutions, a scheme administered by the Higher Education Authority.

DCU has, for many years, been involved in sensor research, including chemical and bio-sensors, says MacCraith. "Sensor research has been going on here for 15 or 20 years. These are systems and devices for making very precise measurements in a range of applications. Increasingly, the focus is on biomedical applications." The NCSR is always developing other forms of intelligent food packaging. In particular, it has developed packs for monitoring fish freshness based on the gas emitted when fish degrades. This work is carried out in the lab of Dr Gillian McMahon and Prof. Dermot Diamond.

The Intellipak team developed a special film that could be built into the packaging and that had special characteristics. It has tiny micropores that can hold luminescent dyes, which means it can be printed upon and is inexpensive. Prof. Han Vos, at DCU, developed dyes that have measurable levels of fluorescence, depending on either oxygen or carbon dioxide levels, and team members Dr Aisling McEvoy and Dr Colette McDonagh worked on developing the special film.

The dye is applied to the film in barcode-like stripes on the inside of the pack. An inexpensive opto-electronic scanner can read the fluorescence of the barcode stripes, thus showing whether the MAP gas mix has changed. It is accurate to within a fraction of one percent, says MacCraith. The luminescent dyes don't react with the gases, only indicate their levels, so it is a reversible, non-consuming interaction, he adds. The approach allows flow-through scanning, so virtually every pack can be tested for integrity and only those that fail need be repacked or discarded, saving unnecessary loss.

It would be simple enough for a consumer to use a scanner at home to confirm product safety, he says. "This is going to add very little to the cost of the packaging but it gives huge added value," MacCraith believes. "There are some similar systems around the world, but we have shown we can test for two gases with one scanner," he says. "We believe that there is massive potential business for this. It really is a multi-billion market and it is an Irish solution."

The team is in the process of patenting the system and has involved a venture capital company to work up a business plan.

Thursday, 30 January 2003

🞣 **DCU National Centre for Sensor Research:**
http://www.ncsr.ie/index_home.html

Concrete and wet tissue

A team in Cork hopes to replace lost vision by implanting microchips in people's eyes.

Electronics specialists and eye surgeons in Cork have joined forces to find ways of reversing progressive blindness. A new research project funded by the Irish charity Fighting Blindness hopes to use microchips that work as virtual eyes, linking with cells in the optic nerve to return some level of visual response.

Retinitis pigmentosa and age-related macular degeneration are the two leading causes of progressive blindness. Both involve the loss of rods and cones, two cell types in the retina that are essential for vision. Research groups around the world are attempting to develop electronic replacements for the rods and cones, according to Dr Anthony Morrissey of the biomedical microsystems group at the National Microelectronics Research Centre, part of University College Cork. The idea is to develop microelectrode arrays that can send signals into the optic nerve, replacing those sent by the rods and cones. "The microelectrode array would do the job of those cells."

It is still unrealistic to suggest that anything like proper sight might be achieved, says Morrissey. The systems could, however, give back enough vision to restore a sense of night and day, something Fighting Blindness says would be a major advance. In time, as the devices improved, their users might be able to identify large objects, such as cars or open doors, says Morrissey.

The centre has decades of experience of developing advanced microelectronic devices. The biomedical microsystems group was set up about four years ago to look at implants and biosensors. It involves a team of 12–14 researchers, including electronic engineers, chemists, biochemists, microbiologists and physicists. Fighting Blindness decided to help the centre get involved in world research efforts in this area. Its three-year grant is "like a foot-in-the-door project", says Morrissey. It allows the centre's team to apply its expertise to the challenge.

He and colleague John Alderman joined with surgeons Prof. Philip Cleary and Dr Hossein Ameri of Cork

University Hospital to develop devices and surgical techniques to provide the electronic vision. The technical and the surgical challenges are enormous, says Morrissey. "The surgeons say attaching a chip to the retina is like trying to attach a concrete block to a wet piece of tissue paper." The devices must be very small and very thin but able to accept a signal, then redistribute it across a microelectrode array to produce 'vision' that can be interpreted by the brain.

"We have started to move towards flexible implants based on polyimide film,"" says Morrissey. "It is tricky to make them small and flexible enough, and one of the major issues is getting them working." The team is building a polymer film complete with electrodes—squares of platinum—that can be pressed against nerve cells. This can be built into layers, creating more complicated circuitry. To this can be added microcoils, which allow low voltages to be induced in the circuits, and a 'microplexer' that can get more information out of a single incoming signal.

The centre's team is also modelling these devices, simulating the circuitry and assessing how it might perform when in position. In addition, it is studying the effect an implant has on eye tissue and the problems it can cause. The hospital team is evaluating the best surgical approach—'epiretinal' placement, putting the chip on the surface of the retina, or 'subretinal' placement, putting it just under the retinal tissues. "Each has its own advantages and disadvantages," says Morrissey, but the UCC team prefers subretinal. "It is less difficult to slot an implant in behind the retina than having to tack it onto the front."

The Cork group has established good links with the world's leading researchers in the area, including Dr Mark Humayun in Los Angeles and Prof. Rolf Eckmiller in Cologne. This should help move the research along more quickly, Morrissey believes, because everyone involved wants the same thing: a way to restore lost sight.

"There isn't international competition. People are trying to help each other," says Morrissey. "The more people there are looking at this problem, the sooner it will be solved."

Thursday, 13 February 2003

+ **National Microelectronics Research Centre** (now part of the Tyndall National Institute): http://www.tyndall.ie/

Cork University Hospital: http://www.shb.ie/class-17674347.cfm

Pulsar discovery at NUI Galway

Astronomers have
made an important
new discovery about
the mysterious pulsar.

The Irish climate was no impediment to a group of Irish astronomers, which for the first time has answered important questions about the enigmatic astral body, the pulsar. These objects have defied attempts to understand them since their discovery in 1967, but the NUI Galway group of astronomers has revealed at least some of their secrets.

Pulsars are a most unusual kind of star that emits powerful pulses of radio waves. Ironically, the Irish were involved in pulsar study from the very beginning via the great Irish astronomer, Jocelyn Bell, who discovered them. The NUI team, which is working in collaboration with a group in Amsterdam, measured for the first time a link between the intensity of optical light coming from the pulsar and the intensity of its radio wave emissions, publishing its results two weeks ago in the journal, *Science*.

The finding could reveal significant new information about how these mysterious bodies work, explains NUI Galway lecturer, Dr Andy Shearer, who led the Irish research group. "We had been working on observations of pulsars for the last 10 years. One of the unanswered questions was whether there was a link between the optical signal and the radio signal."

Pulsars only arise at the end of a star's lifetime. Large stars 'die' in a massive supernova explosion and then collapse inwards to form what is called a neutron star. They become extremely dense in this state with all the mass of a star similar to our sun compacting into a space no more than 10km across, less than the distance across Dublin. A mere teaspoon of this super-massive material would weigh more than a billion tonnes, and neutron stars rotate very rapidly, up to 600 revolutions per second.

Young neutron stars, pulsars, emit a powerful radio frequency signal that beams out like the light beacon of a lighthouse, with the signal linked to how fast it rotates, explains Shearer. A very few, however, also give off visible light. "About 1,400 pulsars are known, most of them seen as radio objects," he explains. "Only eight are observed to pulse optically. Understanding the pulsar phenomena remains one of the unsolved problems in astrophysics."

The team decided to answer at least one question, whether there was a link between the intensity of the radio signal and visible signal coming from the handful of pulsars that gave off observable light. They turned their attentions to the 'Crab' pulsar in the Crab Nebula, which rotates 33 times a second.

The Galway group over the past decade had developed a unique world-class camera system known as TRIFFID. Built using funding from Enterprise Ireland, TRIFFID was able to capture the weak visual signal coming from the pulsar when attached to the William Herschel Telescope in La Palma. The matched radio signals were recorded using the National Radio Observatory at Westerbork run by the University of Amsterdam. "We observed over 10,000 giant radio pulses and discovered, for the first time, that there is a link between the radio and optical signals from pulsars," says Shearer.

"The Crab is very unusual in that the pulsed radio emissions vary. About once every second the radio pulse jumps in intensity to 1,000 times its usual strength. The team was able to match this with a 3 per cent increase in optical signal intensity. This is the first time these types of observations have been seen," he said. "When we did the calculations, we found the amount of energy released is about the same," this despite the change in intensity, he added. "That is what is difficult to explain."

NUI Galway and other groups around the world are now trying to understand the meaning of the new findings and to discover where on the neutron star the radio and optical signals come from, for example the surface rather than the interior. While the discovery raises as many questions as it answers, it is extremely valuable, Shearer believes. "Our observations have, for the first time, linked emission from these two parts of the electromagnetic spectrum, and in doing so ruled out some of the competing models for pulsars."

Thursday, 07 August 2003

⊞ **NUI Galway Computational Astrophysics Laboratory:** http://cal.it.nuigalway.ie/
Westerbork Observatory: http://www.astron.nl/p/observing.htm

A. **Shearer,** B. Stappers, P. O'Connor, A. Golden, R. Storm, M. Redfern and O. Ryan, (July 2003) Enhanced optical emission during Crab giant radio pulses, *Science* 301, 493–95.
Science: http://www.sciencemag.org/cgi/content/abstract/301/5632/493

+ Crab Nebula

+ Dr Geraldine Butler

A new way to protect premature babies

A scientist is studying an emergent yeast species that has become a particular danger to infants.

A common but little-known yeast is a real danger to newborn and premature babies because of its ability to form a thin coating known as a biofilm. Now, a researcher at University College Dublin using advanced genetic technologies hopes to find the yeast genes responsible, which could produce information that would lead to new kinds of treatments.

"The biofilms are drug-resistant and are a particular problem for premature neonates," says Dr Geraldine

Butler, senior lecturer in UCD's Department of Biochemistry. Also involved in UCD's Conway Institute of Biomolecular and Biomedical Research, Butler recently won a four-year Science Foundation Ireland investigator award worth almost €1 million. She will use this to study a pathogenic yeast, Candida parapsilosis, an 'emergent' yeast that causes an increasing number of infections.

C parapsilosis is related to the more familiar C albicans, but although C albicans has come in for extensive study and its genome has been sequenced, very little is known about C parapsilosis, despite its new importance as a cause of disease, says Butler. It is responsible for half of all infant infections and a quarter of general infections associated with the candida organism, she says. "C parapsilosis is not sequenced at all. There is almost no sequence information for it."

She will use her award to correct that situation, carrying out a 'genome sequence survey' of C parapsilosis. "It is not a full sequence, it is a survey of the genome," she says. "What I am trying to get is partial information of a lot of genes. It will tell us quite a lot about the genome structure." The full genome has between 26 and 30 million steps, or base-pairs, and an estimated 6,500 genes. Butler's survey will provide a rapid overview of the yeast's genome and help identify where the genes are without a full sequence. "I would hope to identify a third of them," she says.

Her targets are the genes linked to the yeast's ability to form biofilms, colonies of individual yeasts that join together to coat medical implants, catheters and other devices. They are resistant to drugs in this form and often cause blood infections that force the removal of what might be a life-sustaining implant.

The survey will involve chopping up the yeast's genome into about 5,000 fragments. The team will then sequence only 400 to 500 bases at either end of the fragments, data that help identify the genes quickly. The next step is to find genes that are associated with biofilms.

"What I really want to do is try and understand why this organism makes biofilms, because this is why it is a problem organism," she explains. "No one really knows what makes cells develop in this way. We are trying to find out which genes are switched on to form films rather than when the organism is free-living."

She and team members Dr Mary Kelly and Sean Laffey will prepare an 'expression pattern' for genes, comparing biofilm and free-living yeasts to see which genes are on or off. They will use micro arrays that can confirm the presence or absence of thousands of proteins, the substances produced by the genes, in a single sample.

This in turn will give a wealth of information about the genes linked to the formation of biofilms, Butler says. A better understanding of the process should throw up new ways to treat the biofilms and prevent infections without having to remove the medical implants.

Thursday, 04 September 2003

UCD Department of Biochemistry: http://www.ucd.ie/biochem/
Conway Institute: http://www.ucd.ie/conway/

Biomedical research moves up a gear

Patient care should benefit greatly from advanced research at the Conway Institute of Biomolecular and Biomedical Research. The knowledge gained from detailed work in the laboratory can quickly be used to improve their treatment.

A new centre will help link basic medical research with better treatments for patients.

A good example involves investigations by Professor Catherine Godson, an associate professor of medicine at University College Dublin and the lead co-ordinator for molecular-medicine research at the Conway. The work looks at how diabetes damages the tissues of the kidneys and how a substance produced by the body can help to reduce harmful inflammation. Both projects involve examining what goes on inside cells and how they respond to disease conditions.

Godson explains that diabetes has become an epidemic, affecting increasing numbers of both adults and children. With the disease comes long-term damage to a variety of tissues, including those in the kidney, the retina and the smallest blood vessels. Godson and her team are looking at diabetes-related kidney-cell death, called nephropathy.

Kidney-cell cultures are exposed to the high sugar concentrations typical of diabetes, a disease that stops the body from properly controlling blood-sugar levels. The cells respond by releasing proteins; the team identifies and characterises the substances associated with nephropathy. Discoveries so far include completely novel proteins and

proteins whose link with nephropathy had been unknown. The work could lead to new treatments to block the cell death caused by nephropathy, and the research could lead to diagnostic tests to recognise the presence of marker proteins long before the subsequent damage becomes detectable.

Another Conway project involves a molecular study of lipoxins, substances produced by the body to halt damaging inflammation. The body responds vigorously to fight off any infection, producing the effect we recognise as inflammation. Many diseases, including arthritis, lupus and psoriasis, occur when the body doesn't reverse the inflammatory response, producing a long-

+ Professor Catherine Godson and Dr Madeline Murphy

term chronic inflammatory reaction that causes lasting damage and discomfort. "The reason we are interested in them [lipoxins] is they have anti-inflammatory properties," says Godson, "and they aren't just anti-something, they help other cells to shut down inflammation." Her research has shown that the presence of lipoxins can boost the clean-up activity of a type of white blood cell, the macrophage. These cells swallow up and remove dead and dying cells in a process called phagocytosis. It is an essential service, because dead cells can provoke inflammation if they are left behind.

In work funded by the Health Research Board, the Wellcome Trust and Enterprise Ireland, she and colleagues conducted a series of *in vitro* and *in vivo* experiments showing that phagocytosis activity increases between 30 and 50 per cent in the presence of lipoxins. The work is important because inflammation is usually treated with painkillers or sometimes steroids, drugs that have powerful side-effects. Understanding the molecular processes of phagocytosis and the part played by lipoxins could provide alternatives to steroids to treat chronic inflammation, says Godson.

Thursday, 11 September 2003

+ **UCD Conway Institute:**
http://www.ucd.ie/conway/

Medical science merges man and machine

The TV programme *The Six Million Dollar Man* popularised the notion of a mechanised human. Now engineers and medical researchers are joining forces to make science fiction a reality.

Bioengineering is the new frontier, according to the winner of the Royal Irish Academy Parsons Award.

The winner of the Royal Irish Academy Parsons Award in Engineering Sciences 2003 last night talked about the brave new world of bioengineering, research that blends man and machine. Bioengineering has already had a major impact, with the implantation of replacement hip, shoulder and knee joints and the use of cardiovascular stents all now common procedures.

Professor Patrick J. Prendergast, assistant professor of mechanical engineering at Trinity College Dublin, last night delivered his Parsons Award talk at the RIA, in Dublin, entitled *'Life and Limb: The Bioengineering of Prostheses and Implants'*. Siemens sponsored the lecture.

Prof. Prendergast is also the director of the Trinity Centre for Bioengineering, a research unit involving engineering and health science researchers set up with funding from the Higher Education Authority-organised Programme for Research in Third-Level Institutions.

"The idea of the human body as a machine originated with Descartes, like so many ideas in mechanics," Prof. Prendergast said. Although Descartes presented the ideas, the first book on the subject was by an Italian, Giovanni Borelli (1608–79), who sought to extend into biology the rigorous analytical and geometrical method developed by Galileo in the field of mechanics.

"He made some important discoveries, including how high the forces are acting at the joints,"Prof. Prendergast said. "This is what people who do bioengineering are interested in." Bioengineering is about the development of biomedical devices, Prof. Prendergast told his audience. He discussed four different types of devices, starting with replacement hip joints. Surgeons tend to select 'off-the-shelf' hips but there are currently about 500 different designs, he said. "Our research is trying to

+ Professor Patrick J. Prendergast

that raises issues related to "the design of the human body itself and how it comes to have what it has."

He described theories of 'mechano-regulation rules', the way the body's tissues adapted to the mechanical stresses and strains of living. How mechanical stress affects tissues could be seen in bone growth, fracture healing and loss of bone density with ageing. If there was too much mechanical loading of a broken bone it wouldn't set, but too little and it won't heal, Prof. Prendergast said. He was researching the optimal amount of 'mechanical stimulation' to assist healing, something that might also help bone materialisation to counter osteoporosis right through the body.

"To keep a healthy skeletal system we must stimulate it with mechanical loads," he said. "We have to get out of our minds this idea that the skeleton is a fixed structure." He illustrated this particular point by describing the differences between flighted ducks and flightless domesticated ducks. "Flightless ducks have got a different bone structure than those that fly," he said. Size and strength are optimised for flight, but ground dwelling ducks have heavier, inefficient bones that preclude them from flight.

show which of these is best and how to select the right one for the patient."

He talked about replacement shoulder joints. "The shoulder is one of the most difficult to replace," Prof. Prendergast said. His research group has developed several new designs in an ongoing effort to improve this replacement. His team is also studying cardiovascular stents, a tube-shaped wire mesh implant used to keep arteries open. It has joined US manufacturer Medtronic AVE Ireland to develop new stent designs.

The most intriguing implants are those for the three bones of the middle ear that control hearing, said Prof. Prendergast. All mammals, including humans, have three bones to control hearing but all other species have just one, he said, something

This added a dimension to evolution, he concluded. "Even Darwin could see that [evolutionary pressures] were selecting species but also how the species would grow."

Thursday, 09 October 2003

⊞ **Trinity Centre for Bioengineering, TCD:** http://www.tcd.ie/bioengineering/index.php
Medtronic Ireland: http://www.medtronic.com

NI professor joins line of distinguished researchers

Film clips on your mobile, wireless internet on your PC and better sound and pictures on your DVD are results of the research done by John McCanny.

Professor John McCanny wants your DVD to be as sharp as possible. He also wants your CD player to deliver music that sounds as good as the original. He even wants your mobile phone to be able to play movies. All of these innovations flow from the advanced research he directs at Queen's University. It is all about digital signal processing, enabling digital sound and video devices to handle masses of information as fast and efficiently as possible, while gobbling up very little energy.

Prof. McCanny, professor of microelectronics engineering at Queen's, this weekend won the 2003 Royal Dublin Society/Irish Times Boyle Medal and Bursary. The award recognises excellence in scientific research and has been given over the past century to some of Ireland's leading scientists. Prof. McCanny is the latest in a long line of distinguished researchers to receive the award, which is named after the father of modern chemistry, Sir Robert Boyle.

McCanny has been researching ways of improving digital signals for more than 20 years. A graduate of the University of Ulster, he is a computational physicist who developed a way to blend microchip technology and advanced mathematics to help polish and speed up digital signals passing through electronic devices.

"It is essentially doing mathematics on silicon," Prof. McCanny explains. "The area is called digital and image processing. What we are trying to do is take mathematical computations you might have done with programming before and put hard-wired computational structures on chips," he explains. "It allows you to do the maths in a highly efficient way. The advantage is it is 1,000 times more efficient than using programming. It means either blazing speed or very low use of power."

Speed and low power use are the two key advantages coming from Prof. McCanny's designs. It means that quite small devices, such as your mobile phone, can handle more data in a shorter time and without draining away all the battery power. This would allow short video clips to play on the mobile phone in your pocket.

He suggested ways that this technology might be used in the near future. He pictures a business commuter waiting for a flight to depart. He or she flips open their laptop computer and switches onto wireless internet, but also decides to watch a film over the net or run a DVD locally. The laptop would support all these services on a single screen, thanks to the ability to handle massive amounts of data quickly while depending only on a small, rechargeable battery.

"We are creating dedicated silicon architectures and building chips that exploit advanced mathematical algorithms. The challenge of all this is how to do it," he says. "All of these kinds of things, people take for granted you can do it. But it takes lots of computer power." His algorithm/chip combinations give mobiles, televisions and video recorders the kind of processing capacities that will enable these new services.

His latest project is a £35-million sterling (€49-million) development under way on Queen's Island in Belfast, part of the old Harland and Wolff complex. Called the Institute of Electronics, Communications and Information Technology, it will be based in the new Northern Ireland Science Park in what is known as the 'Titanic Quarter', part of the harbour adjacent to the dry docks where *HMS Titanic* was built.

"This is the flagship project on the Park," says Prof. McCanny. "The basis of it is we

are trying to develop a new environment of innovation and company creation as part of the new knowledge economy. The intention is to be a Stanford or MIT research centre, working with blue-skies technologies and creating new companies."

+ Professor John McCanny

He plans to use the Boyle Medal Bursary to bring in a graduate student to work on his newest research area, data encryption. He is developing new combined algorithm/hardware devices that will improve the security of data moving across the internet and new wireless communications services.

Monday, 13 October 2003

➕ **QUB School of Electrical and Electronic Engineering:** http://www.ee.qub.ac.uk//
Institute of Electronics, Communications and Information Technology: http://www.ecit.qub.ac.uk/

Irish expert monitoring in-flight radiation

An astrophysicist with the Dublin Institute for Advanced Studies plays an international role in assessing radiation risks to aircraft crew and astronauts.

Ireland has no formal space programme but an Irish physicist has developed a key role in assessing the safety of space travel. His experimental data has also helped set EU standards for radiation exposure limits to aircraft flight crews.

Prof. Denis O'Sullivan has been putting experiments on board NASA spacecraft for 35 years and for more than a decade has measured the radiation risks faced by pilots and cabin crew on board conventional aircraft. An astrophysicist at the Dublin Institute for Advanced Studies, he currently leads an EU research programme in air crew safety involving seven European laboratories and last week saw his ninth experiment being carried aloft into space.

All of us receive a radiation dose from the sun and also cosmic rays coming from deep space. This radiation dose increases the more time we spend on aircraft and the further from the equator we fly. The urgent need to have a clear understanding of the radiation levels involved were seen in research published yesterday in the British Medical Journal title, *Occupational and Environmental Medicine*. It carried a number of Scandinavian studies indicating flight crews carried an increased risk of contracting breast and skin cancer.

One Icelandic study showed a five-fold increased risk of breast cancer. And while a separate Swedish study didn't establish an increased breast cancer risk, it found an increased risk of malignant melanoma among both male and female cabin crew and an increased risk of other skin cancers among the men.

O'Sullivan, with colleague, Dr Dazhuang Zhou, is co-ordinating an international team looking at cabin crew radiation exposures. "We have been flying detectors on everything from the Concord down to the Irish government jet," he said. Airlines around the world have co-operated in this activity. "It is telling us the radiation exposure increases with altitude as we might have guessed."

It also confirmed that the further from the equator we fly the greater the exposure.

In 1990 the International Commission on Radiological Protection recommended that exposure of aircrew to cosmic radiation should be considered occupational exposure. The legal consequences of this were that the EU brought in legislation recently to protect aircrew, a decision reached after an assessment of O'Sullivan's data. He has spent years studying astronaut exposure in spacecraft and in the International Space Station (ISS). His ninth experiment, a collaboration with a Belgian group, reached the ISS after a Soyuz lift-off just last Saturday.

The institute team is providing cosmic ray exposure measurements on a colony of bacteria supplied by the Belgians. The areas of particular interest are bacterial genetic stability and DNA rearrangement after exposure. There is growing interest in what happens during prolonged space flights to bacteria that travel with the crew. "We were invited to take part in the American/Russian mission to the ISS," he said. "Our job is to measure the actual amount of radiation that hits the bacteria and the microbiologists in Belgium will assess the impact on the bacteria." Radiation can cause mutations in DNA, changes that might alter the characteristics of a given organism.

Changed organisms could present an unexpected danger to crew or cause damage to materials, explains O'Sullivan. He has collaborated with NASA for 35 years. "I actually worked with the lunar samples that Neil Armstrong brought back with him on the Apollo 11 flight to the moon," he says.

Thursday, 23 October 2003

⊞ **Dublin Institute for Advanced Studies School of Astrophysics:**
http://www.dias.ie/index.php?section=cosmic&subsection=index&school=astrophysics

A Linnersjö, N Hammar, B-G Dammström, M Johansson and H Eliasch, (November 2003) Cancer incidence in airline cabin crew: experience from Sweden, *Occupational and Environmental Medicine* 60, 810–14.

Occupational and Environmental Medicine: http://oem.bmjjournals.com/cgi/content/abstract/60/11/810

+ Russian Soyuz TMA-3 rocket is transported
  to its launch pad in the early morning

+ Picture of trapped rubidium atoms
cooled to a fraction above absolute
zero, taken by a high-speed
Andor camera

Youth Science Week inspired entrepreneur

It is a special kind of camera that is just as effective taking images of objects 10 millionths of a metre across or 500 light years across. Yet such a camera exists, invented by a Belfast company that grew out of advanced physics research.

Andor Technology Ltd was set up in 1989 as a 'spin-out' commercialisation from Queen's University's Department of Physics, says the firm's managing director, Dr Hugh Cormican. "There is a long history in the department of developing diagnostic equipment," he says.

Cormican and company colleagues Dr Mike Pringle and Dr Donal Denvir had been working on diagnostic tools for measuring aspects of their advanced laser research. They needed a method for 'seeing' the laser strike a surface but had no tool for achieving this. "We were developing diagnostics that could capture spectra over

Belfast researchers have developed a new kind of camera that can take pictures of atoms or capture starlight from the far side of the universe.

very short periods of time," he says. They developed a device to do this, a very sensitive imaging device called an electron multiplying charge coupled device (EMCCD), that, because of its sensitivity, was also very fast at capturing an image.

They knew immediately that the new device had good commercial potential and a Queen's campus company was set up to exploit the technology. This grew over the years and eventually moved off campus. It now employs 110 people and has a turnover of about $15 million a year, with 95 per cent of its products destined for export, says Cormican.

The EMCCD technology "basically makes the camera more sensitive", he says. It is sensitive enough to respond to a single photon of light. "The light falls on a sensor, the sensor generates an electronic signal." This in turn is amplified with an electric voltage to boost the signal beyond the level of background noise. The resultant sensitivity means that it can capture images at extremely low light levels and can get well-lit images very quickly. For this reason it is adept at taking weak light signals in astrophysics but also when looking at very small objects in the nano range down to 10 millionths of a metre across.

"A lot of nanotechnology research is being done with these high-end cameras," he says. "It is the world's most sensitive camera." Its sensitivity means you don't have to depend on very long exposures to get an image. "The advantage of this technology is you don't have to wait," he says. One important new use will be to watch proteins and DNA in action, Cormican says, or to watch biochemical processes inside a living cell by filming

them at 120 to 450 frames per second. "One of the future ways of sequencing genes is to have markers for the different bases and you then watch how it is being replicated" using the camera, he believes. It can capture "living cell processes" as they take place in living tissues.

Light sensitivity is also valuable in astrophysics, and Cormican referred to work at University College Cork that makes use of the camera. A team there is studying quasars, which deliver powerful radio frequency emissions as they spin at rates measured in milliseconds. (See the following article in this collection.) Every few revolutions the radio emissions can flare for as yet unknown reasons, and the Cork group wanted a way to see whether there is a matching change in the light intensity during these radio peaks. The EMCCD camera should allow them to see if indeed there is a matching change, something that could help towards a better understanding of these unusual objects, says Cormican. "They are dear but they are worth it," he says of the devices, which can run from $30,000 to $100,000 depending on ancillary equipment.

Surprisingly, Cormician knew as a teenager that he would one day become a technology entrepreneur. He recollected as a 16-year-old coming from the North to attend an Irish Youth Science Week at the Royal Dublin Society. "One of the things that really stuck in my mind was descriptions of the Irish strategy for developing inward investment," he says. "I went home and told all my friends that I was going to found a high-tech company. It was an Irish Youth Science Week that inspired me."

Thursday, 30 October 2003

Andor Technology Ltd: http://www.andor-tech.com/
Queen's University Belfast School of Mathematics and Physics: http://www.qub.ac.uk/mp/

Dazzling deep space discoveries

A Cork Institute of Technology group is leading Irish efforts to learn more about quasars, the brightest objects in the universe.

Think of the brightest sunlight on the clearest day of the year. Now multiply it a million-fold. Such an intense light would still be hopelessly dim compared with the colossal output from a quasar.

Nothing compares to a quasar when it comes to giving off a continuous source of light, explains Dr Niall Smith, a lecturer in the Department of Applied Physics and Instrumentation at Cork Institute of Technology. Our Milky Way galaxy contains about 100,000 million stars; a single quasar can send out up to 10 million times their combined light, lasting a million years. The celebrated gamma-ray bursts can give off more light again but last for mere seconds, so quasars still top the list for continuous light output, says Smith.

He leads five researchers who form part of an EU network of laboratories working towards a better understanding of these enigmatic objects. The institute also has a lead role in Irish involvement in the network, with collaborating researchers at University College Dublin, University College Cork and Tallaght Institute of Technology.

Current theories assume that a super-massive black hole lurks at the heart of a quasar. Material drawn into its unbreakable hold is immediately converted to energy as radiation in wavelengths from radio up to gamma frequencies. Beyond that they are a mystery. "They have a life cycle we don't fully understand," says Smith. "We don't know what is happening in the centre as material flows in and radiation is given off."

Quasars also have a tendency to flare, pulsing out short bursts of extra energy for unexplained reasons. About 10 per cent of all quasars also emit strong jets of radiation. "These are powerful jets. If you were in the line of a jet you would be in trouble."

The European network involves labs looking at the objects at various wavelengths. The Cork group specialises in optical wavelengths; it has some of the best visible-light

+ Calar Alto telescope, Spain

data on quasars available anywhere thanks to a new camera designed by a Belfast company, Andor. (The Andor camera was described in the *Science Today* report on 30 October; see the previous article in this collection.) The team used it at the 2.2-metre telescope at Calar Alto, Spain. Existing devices can take one frame a minute; the new camera can take several frames a second, an important consideration for the data being sought.

Short flares cause the light output of a quasar to jump temporarily. "What we are looking for is variation in the light output of the quasar, both very small variations and very rapid variations. That is why the Andor camera was so important for us." The higher resolution allowed the Cork researchers to capture these very rapid variations in light intensity, which can be plotted along "light curves", explains Smith.
"It allowed us to generate light curves that are very accurate. We found some very nice light curves where we could follow the flare over time. We were looking at a

quasar that has a very powerful jet shining towards us. This type of object allows us to learn more about the physics of what is going on in the jet."

This week the network has been ganging up on a quasar known as 0716+741, the same object studied with the new camera during two runs in January and September at Calar Alto, says Smith. "There is an intensive campaign going on with this one object." It includes 11 radio telescopes, 10 optical telescopes and satellites, all looking at the quasar at different frequencies. "These jets are millions of light years long," explains Smith, and short-term flares can arise along small sections of the jet. "If you see a flare, you can get an idea of the size and get some idea of what is going on."

Thursday, 13 November 2003

⊞ **CIT Department of Applied Physics and Instrumentation:** http://www.physics.cit.ie/

+ Schematic of the central region of a black hole showing infalling
  material (yellow) and a powerful jet of outflowing material (white)

Bugs that cause gut feelings

From "bugs to drugs" is the way University College Cork describes its new €16.5-million Alimentary Pharmabiotic Centre (APC). Dedicated to the study of the bacteria that colonise our digestive systems, discoveries at the centre could lead to new therapies for a range of persistent bowel diseases.

A research centre at University College Cork hopes to discover new drugs by studying the trillions of bugs in our digestive systems.

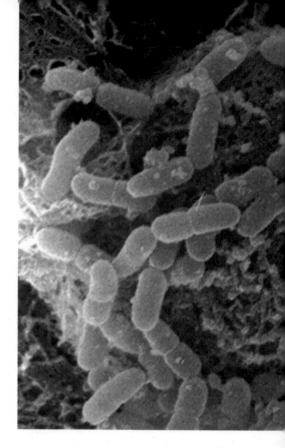

The APC, which officially opened last Friday, is the first of the new large-scale CSETs (Centre for Science, Engineering and Technology) to be funded by Science Foundation Ireland. The CSETs are major undertakings with the largest budgets yet made available for Irish research teams.

The centre at UCC received €16.5 million from SFI, plus funding from UCC-campus company Alimentary Health Ltd and international partner Procter and Gamble. It will examine the role being played by the complex bacterial flora in our digestive tracts, explains Prof. Fergus Shanahan, director of the APC. "In essence we are studying the interaction between bacteria in the gut and the host," says Shanahan, who is also professor and chair of UCC's Department of Medicine.

The *Alien* sci-fi films, where an alien invader takes up residence in a human host, have nothing on the teeming trillions of bugs that occupy our alimentary canals. The bacterial cell count is so high it outstrips the total number of humans that ever lived on this planet. There are more bacterial cells than

there are human cells in the body, adds Shanahan. In effect, the bacterial cells we carry about our person outnumber our own cells. "We are 95 per cent bacterial in terms of overall cell count," says Shanahan. "These bacteria are often described as the hidden organ or the neglected organ of the body."

Taken together the bacterial cells in a single adult can weigh up to two kilos. "It is a living mass that is tantamount to a liver," he says. "When you think of its activity, it is equivalent to an organ. It has to be important." Shanahan, with deputy director Prof. Gerald Fitzgerald of UCC, will now try to discover just how important our bacterial hitch-hikers actually are. Interest in them is not new, but the tools weren't available to really understand their biological significance, says Shanahan. "The difference now is having the technology to study it."

An APC team of 50 scientists and clinicians

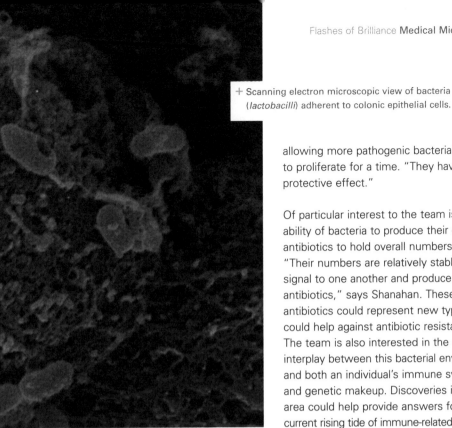

+ Scanning electron microscopic view of bacteria (*lactobacilli*) adherent to colonic epithelial cells.

will gauge the bacterial influence on our gastrointestinal tracts and on our immune systems. Gut flora are essential in kick-starting our immune systems, and without them we would be in trouble, he explains.

Researchers have raised 'clean' mice that lack bacteria in their gut, and invariably they are unhealthy and more susceptible to illness. The bugs also contribute directly to our ability to gain nutritionally from the food we eat. Our calorific intake would have to increase by a third without them, says Shanahan, because they help break down complex molecules. "They are net contributors to your body weight. They are not parasites, they are net contributors."

They can also produce negative effects, he adds. "They are clearly involved in disease, for example antibiotic-linked diarrhoea." Powerful antibiotics can disturb the bacterial balance, knocking out harmless species and allowing more pathogenic bacterial species to proliferate for a time. "They have a protective effect."

Of particular interest to the team is the ability of bacteria to produce their own antibiotics to hold overall numbers in check. "Their numbers are relatively stable. They signal to one another and produce bacterial antibiotics," says Shanahan. These same antibiotics could represent new types that could help against antibiotic resistance. The team is also interested in the complex interplay between this bacterial environment and both an individual's immune system and genetic makeup. Discoveries in this area could help provide answers for the current rising tide of immune-related disorders associated with the gut. Ironically, while both peptic ulcers and gastric cancer are in decline, the numbers of patients presenting with Crohn's disease, ulcerative colitis, gastroenteritis, food allergy, inflammatory bowel disease, bowel cancer and irritable bowel syndrome are all on the rise.

Some of these may arise from our personal interaction with the gut environment. It may be that alterations in gut flora could provide a trigger that initiates a disease state in the bowel.

Thursday, 20 November 2003

+ **UCC Alimentary Pharmabiotic Centre:**
http://apc.ucc.ie/content/
Alimentary Health Ltd:
http://www.alimentaryhealth.ie/

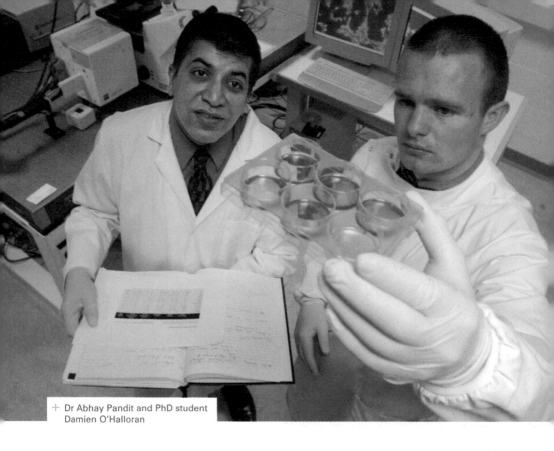

+ Dr Abhay Pandit and PhD student
Damien O'Halloran

New skin hope for those who suffer burns

A Galway research group is copying nature in an effort to produce substitute skin that can promote rapid healing of chronic and acute wounds. It depends on providing a 'scaffold' that the body can recognise and use to grow new tissues.

An NUI Galway research team hopes to develop 'substitute skin', a natural product used to help promote repair after serious injuries.

"It is as good as having your own skin back, but we are trying to fool the body," says Dr Abhay Pandit who heads the research at NUI Galway's National Centre for Biomedical Engineering Science. "We can look at it as a living tissue or something that can produce living tissue in the body." Labs around the world are attempting to achieve the development of a universal second skin that can provide a rapid response to serious burn injuries or non-healing ulcers. "The distinctiveness of our approach is the way we are stabilising the scaffold," he explains.

As with other research groups, the team is using natural proteins to develop a scaffold into which tissues can grow, including fibrinogen and fibrin, associated with blood clotting, and collagen, the body's key connective tissue. Up to a third of all the body's protein content is collagen, so it is readily recognised and tolerated by the body, a key requirement for replacement skin. "The material is something the body can identify. It can relate to it and the cells can understand it," says Pandit. "We are modifying these proteins to tailor their degradation in the body. You want them to degrade because you want the body to produce its own proteins, but the degradation should match the healing rate."

The scaffold in itself is not enough to promote healing, states Pandit, who is attached to NUI Galway's Department of Mechanical and Biomedical Engineering. "The proteins act as a scaffold, then we add in other factors to that. We need to give extra factors to promote healing. The beauty of the scaffold is you can tailor the factors we use." This requires a high degree of interdisciplinary co-operation, hence the idea of basing the research project in the Centre. "The research programme will bring together the fields of gene therapy, biomaterials and biochemistry into the therapeutic domain." Separate groups can look at the scaffold construct while others seek protein targets that promote tissue growth.

"We are actually trying to achieve a dual approach," he says. The scaffold provides immediate protection, say after a burn injury, and is accepted by the body. The extra protein factors it carries then encourage the body to heal and grow replacement tissues. This quick response is essential if the approach is to be used in the event of a burn. A slightly different approach is needed in the treatment, for example, of non-healing leg ulcers, a common condition in diabetics. Ulcers account for about 50 per cent of all lower-limb amputations in a given year and a study in Waterford several years ago showed that 11 per cent of the patients with diabetes had experienced non-healing leg ulcers.

The team's main goal is to produce a bilayered composite, with one layer that mimics the skin and another that directs dermal regeneration. "Using analogues of human skin, this research is geared towards producing a final product which would reduce the need for skin grafts and would also result in significantly less suffering being endured by the patient." While skin replacement is the initial target, the same approach could be used in a wide range of tissues, Pandit adds, from cartilage to organs. The key is to have a scaffold and embedded factors to match processes natural to the body.

Funders for the research include Enterprise Ireland, the Health Research Board, the Irish Research Council for Science, Engineering and Technology and the Centre itself. If successful, the research could "lead to clinical trials for treatment of chronic wounds and investigative work in tissue engineering" that could benefit millions of patients worldwide, says Pandit.

Thursday, 11 December 2003

NUI Galway National Centre for Biomedical Engineering Science:
http://www.nuigalway.ie/ncbes/
NUI Galway Department of Mechanical and Biomedical Engineering:
http://www.nuigalway.ie/mechbio/

Condition creates a confusion of senses

The term used to describe this is synaesthesia, and a multidisciplinary team of Irish researchers has begun a new study to determine how many people here have it. It runs in families and women are three times more likely to have it than men, according to Dr Kevin Mitchell of Trinity College Dublin who is heading the project.

When your eyes scan a page of text do you see certain letters in colour? Can you taste colours or sounds? If so, you might have an unusual condition in which sensory signals get mixed up to deliver a unique view of the world.

As many as one in 2,000 people exhibit synaesthesia to varying degrees, says Mitchell, a lecturer in Trinity's Smurfit Institute of Genetics. "We want to see how common this is in the Irish population and see what the genetics of it is," he says.

Partners in the research include Dr Fiona Newell, a cognitive scientist in Trinity's Department of Psychology, Dr Aiden Corvin, of the Department of Genetics and Ciara Finucane, also of the Department of Psychology. Mitchell himself is a specialist in developmental neurobiology or neurogenetics. "We are interested in the genes that specify how the brain should wire up," he says.

The brain is a staggeringly complex organ and researchers are only beginning to understand how the brain's various circuits are connected. "The basic thing we are trying to understand here is how different parts of the brain link to different functions," says Mitchell. "During development those areas have to be all wired up correctly."

The normal wiring pattern goes somewhat awry for those with synaesthesia, he says. "It can best be described as a mixing of the senses." There might be an audible stimulation but the person hears as well as 'sees' the sound. "Most commonly the person experiences the perception of colour in response to a non-visual stimulus," he explains. Many synaesthetes also have extra colour perception when looking at plain black text.

"They will see particular letters of the alphabet or numbers in a particular colour. That colour will be particular to that individual," he says. "The number five might always be red, the number two might always be yellow. The question is, how is this happening? It relates to how we integrate perceptual information from the various senses."

Sensory inputs from the eyes should only reach visual centres in the brain, but extra

PQRST UVWXYZ
'89

5 5 5 5 5 5
5 5 5 2 5 5
5 5 5 2 5 2 5
5 5 2 2 2 5
5 5 5 5 5 5 5

+ Synaesthasia can cause some people
to see letters and numbers in colour.
Pictured on the left the numbers 2 and 5
in black and white, as might be seen by
most people; on the right as they may be
seen by those with synaethesia.

connections seem to exist that link say to hearing or colour or taste. "We think there is some cross-wiring happening. We are not sure why this happens." Generally it is harmless, but if severe it can be very disturbing, with say a touch producing a cascade of sensory responses including smells, sounds and colours. "In some rarer cases it can be highly distracting," says Mitchell.
While there is some data suggesting synaesthetes have more difficulty with maths, it can assist learning. "Things are naturally colour-coded for them," he says. "It also seems to be more common among artists and composers," with Franz Liszt, Rimsky-Korsakov and Wassily Kandinsky all thought to be synaesthetes.

Although potentially quite common, few are aware of it. "One of the reasons some people haven't heard of it is if you report as a child you tasted colours or smelt sounds you would be corrected by parents or made fun of by other children," says Mitchell.

"We want to look at whether people have multiple forms of synaesthesia, and also whether a particular kind of synaesthesia runs in a family or if the family members have different forms." It is three times as common in women, suggesting it might have something to do with the X chromosome. "Or maybe there is something different in the way the male and female brain is wired that makes women more susceptible to it," says Mitchell.

"We would like people to contact us and fill in a questionnaire that will give us some details of the kind of experiences they have had." People may then be asked if they would be willing to undertake simple visual tests in order to measure how strongly they experience synaesthesia.

The ultimate goal over time would be to do a genetic study of the genes associated with synaesthesia, given the permission of those involved, says Mitchell. Understanding how the brain sometimes miswires itself could lead to a better understanding of how more typical brain connections occur.

Thursday, 11 December 2003

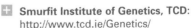

Smurfit Institute of Genetics, TCD:
http://www.tcd.ie/Genetics/
TCD Department of Psychology:
http://www.tcd.ie/Psychology

Scientists search for a better MMR jab

A Trinity-based virologist has received substantial new research funding to develop a measles, mumps and rubella vaccine with fewer side effects.

A dangerously large number of Irish parents have resisted repeated calls to have their children vaccinated against measles because of fears that the jab can cause serious side effects. Now a Dublin-based research team plans to develop a safer vaccine that might overcome these fears.

It is wrong for advocates of the three-in-one MMR vaccine to dismiss the fears as unfounded, argues Prof. Greg Atkins, a virologist and director of the Moyne Institute of Preventative Medicine at Trinity College in Dublin. "The MMR vaccine project stems from the fact that there have been a number of adverse effects related to the vaccine," he says.

Autism is the biggest fear for parents, and despite the fact that repeated large-scale studies suggest the autism –MMR link does not exist, there are other side effects that make the development of a safer vaccine highly desirable, says Atkins.

The current MMR vaccine uses three live, attenuated— or weakened—viruses for measles, mumps and rubella. "The adverse effects come from the fact that they all use live viruses," says Atkins. Although the MMR vaccine currently in use does pose risks, he remains strongly in favour of vaccination. "The evidence of adverse effects for the wild-type virus are much worse, a thousand times or more than the vaccine," Atkins acknowledges. "Certainly you should be vaccinated, but there are adverse effects, which are well established, and nobody argues with them. Medical people tend to ignore them. What we are trying to get away from is the use of live viruses, because this has produced the risk."

To this end, Atkins and his team recently received a Science Foundation Ireland programme grant worth almost €700,000 for the development of a safer recombinant MMR vaccine. "This is part of a research programme that is now worth over €2 million that uses viral vaccines to treat human and animal diseases." His approach is to use viral RNA to help stimulate a powerful and lasting immunity against the three diseases. "We are using recombinant RNA molecules. This isn't a DNA vaccine," he explains. "We are developing new technology to use naked RNA molecules as a vector to express the proteins produced by the measles, mumps and rubella viruses."

DNA resides in the nucleus and a DNA-based vaccine would have to carry out its action there, getting into the host genome and replicating itself to spark immunity. This it does by releasing messenger RNA that works inside the cell but outside the nucleus to produce viral proteins. The Trinity group's approach is to use RNA instead. "It will not insinuate itself into the genome so it can't replicate," says Atkins. The recombinant messenger RNA is still capable of producing the same proteins as the live virus but has the added advantage of only achieving one round of protein production before breaking down.

The real challenge, however, is finding a way to get the RNA into enough cells to induce a sufficiently high level of immunity. Regular injection wouldn't reach enough cells and the RNA must be kept safely within a cell to do its work. It degrades rapidly outside the cell and "would be completely useless", Atkins says.

The National Microelectronics Research Centre at University College Cork has come up with an answer to this problem, known as 'electroporation'. "The electroporation is a new way of getting the RNA into the cells," says Atkins. It involves delivering a very low-level electric shock at the point of injection. It is so small as to go unnoticed by the person receiving it, says Atkins, but it causes the cells briefly to become more

porous, allowing the recombinant RNA to slip inside.

Once there, the messenger RNA performs more or less the same function as RNA released from the cell nucleus. The viral RNA is translated into a viral protein, a substance that the immune system recognises as foreign. Atkins plans to include seven different RNA segments to spread the immune response across different forms of the three disease-causing organisms.

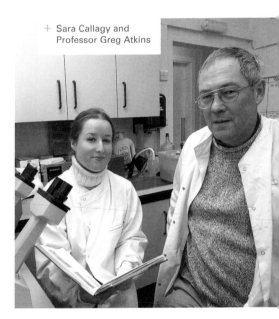

+ Sara Callagy and Professor Greg Atkins

If everything works, the research should lead to a much safer vaccine with fewer side effects. "This work is proof of principle," adds Atkins, but it will be a number of years before all the tests can be done to confirm safety.

Thursday, 18 December 2003

⊞ **Moyne Institute of Preventive Medicine, TCD:** http://www.tcd.ie/Microbiology/index.html

Impact of diabetes on kidneys to be probed

New research at UCD into diabetes could help deliver improved treatments for Alzheimer's and other neurodegenerative diseases.

About 100,000 people in this country have diabetes and know it. Another 100,000 probably have this life-threatening disease but are unaware they are at risk. The rapid growth in the numbers presenting with this disease has taken the medical community by surprise. It has become a common hazard of our 21st century lifestyle and our new-found tendency to be overweight.

The disease occurs when the body fails to regulate blood-sugar levels properly. This happens because either the body doesn't produce enough insulin, the hormone responsible for controlling sugar levels, or the body can't make proper use of the insulin that is available. Either way, high blood-sugar levels play havoc over time with the body's sensitive systems. It does serious damage to the retina, to blood circulation in the limbs and also to the kidneys, so much so that a kidney transplant may represent the ultimate treatment.

Dr Derek Brazil of University College Dublin's Department of Medicine has started a new research project focusing on damage caused by diabetes to the

+ Rosemarie Carew, Dr Sarah Roxburgh, Dr Derek Brazil and Kristine Nyhan

kidney, a condition known as diabetic nephropathy (DN). It affects about 30 per cent of all diabetics and he wants to understand why it hits some and not others and whether there is a genetic component to it.

Working from UCD's Conway Institute, he will build on recently published research that showed insulin is important not just in DN but also in protecting normal brain cell development. He and colleagues at the Joslin Diabetes Centre, Harvard Medical School in Boston published their findings last August in the *Journal of Neuroscience*.

"Neurons need a lot of energy," explains Brazil. "It was never known whether insulin had an effect in the brain." The research group studied mice that lacked IRS-2, a key protein essential for insulin production and action in the body. They found that mice with a lack of IRS-2 (and therefore poor insulin function), had 30 per cent smaller brains due to a reduced number of brain cells. They also found that levels of a protein associated with the development of Alzheimer's disease called phospho-Tau were raised in the brains of the diabetic mice.

These results provide strong evidence that insulin and a related hormone called IGF-1 serve a protective function in the brain, according to Dr Brazil's research. They also suggest unsuspected links between diabetes and Alzheimer's disease. "Insulin and the insulin pathway are not only involved in controlling glucose but in the development of the brain itself,'" he says. Dr Brazil hopes to do a similar intensive biochemical study (at the Conway Institute)

of what happens inside kidney cells when DN takes hold. The work is funded by a programme grant from the Health Research Board and from the Conway, which in turn has received substantial support from the Programme for Research in Third-Level Institutions, organised by the Higher Education Authority.

"We are focusing on a complaint of diabetes rather than on the diabetes itself," says Brazil. "We are trying to work out what goes wrong in the kidney with diabetic nephropathy; why does the filtering capacity of the kidney crash? "The first clinical sign of the disease is an enlarged kidney," he says. "It takes 15 to 20 years before the whole thing progresses to the point where you are at end-stage disease," with dialysis and a kidney transplant often the only answer. "We are trying to intervene to prevent the disease or try to delay it."

The DN research team involves 25 to 30 researchers who are delving into cell chemistry to understand the disease in detail. "We are doing genomic analysis of high glucose treated cells from the kidney," he explains. The team is particularly interested in the cell's "signal transduction pathways", how proteins and other substances inside the cell interact. An early target is the insulin/IRS-2 pathway, which proved important in Dr Brazil's earlier work. "It gives you clues as to which pathway can be inhibited to control the disease," he says. "The aim is ultimately to provide clinicians with a better set of targets against the disease."

Thursday, 08 January 2004

⊞ UCD School of Medicine and Medical Science, UCD Conway Institute:
http://www.ucd.ie/conway/

Markus Schubert, Derek P. **Brazil**, Deborah J. Burks, Jake A. Kushner, Jing Ye, Carrie L. Flint, Janet Farhang-Fallah, Pieter Dikkes, Xavier M. Warot, Carlos Rio, Gabriel Corfas and Morris F. White, (August 2003) Insulin receptor substrate-2 deficiency impairs brain growth and promotes Tau phosphorylation, *Journal of Neuroscience* 23, 7084–92

Journal of Neuroscience: http://www.jneurosci.org/cgi/content/abstract/23/18/7084

+ Professor Tim O'Brien

New boost for stem-cell research

NUI Galway's new REMEDI centre brings together a top team in gene therapy and stem-cell research.

Gene therapy meets stem-cell research in a new research institute that should give Ireland a world-class profile in these important research areas. Based at NUI Galway, the Regenerative Medicine Institute (REMEDI) also binds together laboratory research and clinical practice.

This combined approach is evident in the centre's director, Prof. Tim O'Brien, who is professor of medicine in the university and consultant endocrinologist at University College Hospital Galway. "We are attempting to bring together these two technologies, adult stem cells and gene therapy," says Prof. O'Brien, who was recruited from the US Mayo Clinic to head up Galway's drive for the new REMEDI centre. The plan is to use gene therapy techniques to control the growth, development and targeting of the adult stem cells to be used in new clinical treatments likely to flow from the research.

"One of the difficulties is getting adequate numbers of adult stem cells," he says. "We would hope to increase proliferation by increasing gene expression associated with cell growth."

Stem cells offer hope where other surgical and pharmaceutical treatments falter. These are the building block cells, undifferentiated cells with the potential to transform into brain, blood, bone and muscle cells. It is this plentipotentiality that attracts researchers to their potential for new treatments.

The transformation takes place in the presence of specific proteins, produced by teams of genes, says Prof. O'Brien. Understanding this transformation is another aspect of the research to be conducted at REMEDI, which will have a staff of about 30 researchers.

Once the stem cells differentiate, they have to know where to go, where they are needed. The team also seeks the 'homing targets' that signal to the stem cells and bring them to sites of injury, he says. With such information at hand, new treatments become possible for the repair of the damage caused, for example, by Parkinson's and MS.

The centre's stem-cell expertise will be led by Dr Frank Barry. Dr Barry directed arthritis research at Osiris Therapeutics in Baltimore, Maryland and will join the REMEDI team next month.

It is this blend of experience that should enable the centre, which received €15 million from Science Foundation Ireland, and €4 million from industrial partners including Medtronic in Galway and ChondroGene of Toronto, to develop new stem-cell treatments, says Prof. O'Brien.

The SFI funding will build on earlier support provided by the Higher Education Authority-directed Programme for Research in Third-Level Institutions. Its grant of €20 million allowed NUI Galway to build the National Centre for Biomedical Engineering Science. This in turn provides a home for the new REMEDI centre.

Thursday, 22 January 2004

➕ **REMEDI, NUI Galway:** http://www.nuigalway.ie/remedi/

Making a new hip last longer

A Cork engineering student is helping surgeons improve the success of artificial hip-joint operations. Her novel research into a key cause of joint failure has also won her a trip to London next month, and the right to represent Ireland in the annual Institution of Mechanical Engineers' Competition.

Award-winning student mechanical engineer Niamh Thompson is off to London to represent Ireland in an international competition.

Niamh Thompson, a final-year degree student in Cork Institute of Technology's Department of Mechanical and Manufacturing Engineering, has already claimed the national engineering award for the best mechanical engineering degree project. The competition involves all third-level institutions with mechanical engineering degree programmes in the Republic of Ireland. Thompson now goes forward to the international final, competing against the best from Northern Ireland, Wales, England and Scotland. Her focus is the cement used to lock replacement hip joints into position at the top of the leg. "This is the first time this particular research has been conducted," she says.

Hip-joint replacement has become commonplace, with up to a million procedures a year taking place in hospitals around the world. A number of these will fail after a time, forcing more surgery to carry out repairs. Up to 80 per cent of these failures are caused by loosening, either through infection or the failure of the bone cement, says Thompson.

Surgeons from Cork University Hospital and St Mary's Orthopaedic Hospital approached CIT, looking for research into one possible cause for hip replacement failure. "They had a concern that hydrogen peroxide used during surgery was having an effect on the hip joint cement," says Thompson. Surgeons flush the top of the leg bone, the femur, with hydrogen peroxide solution before fixing the metal hip joint replacement in place using bone cement. The assumption was that the solution reacted with either the bone or the cement to increase the risk of later joint failure.

Thompson was looking for a fourth-year research project and had developed an interest in biomedical engineering after a stint working with local artificial knee- and hip-joint manufacturer, Stryker Howmedica Osteonics. She readily volunteered to

+ Engineering award winner Niamh Thompson

peroxide solution. They are tested to failure to establish the influence of the solution on cement breakdown. The segments also undergo microscopic analysis, looking for "crack growth propagation" that works down into the cement from the surface. "The hydrogen peroxide we believe is causing pores, increasing the porosity of the cement. The more pores, the greater the risk of failure," Thompson says.

She has a good working hypothesis of why these pores form. The solution produces an amount of frothing in the bone as the chemical reacts with proteins in the blood. "It is this air in the socket which is causing inclusions in the bone cement. These air gaps cause the bone cement to fail." Much remains to be done. "A lot of the work has been in designing the moulds and in testing the segments correctly," she says. "The testing is still very much in progress. Hopefully I will know by the time I finish up in March if the solution is affecting the cement."

Thompson has always been interested in mechanical engineering. "When everyone else was playing with Barbies I was playing with Meccano," she says. "I have always loved hands-on stuff. That was always the area I was most interested in."

Thursday, 19 February 2004

pursue this research challenge. "The objective of this project is to see how the solution is affecting the bone cement," she explains.

She had to devise ways to test the hydrogen peroxide theory, not just looking at physical changes but also at how the solution might speed up failure due to cracking and fatigue under pressure. She created moulds to form segments of hardened bone cement. The moulds were flushed with the solution before being filled with cement. This simulates what actually happens with the femur before the bone cement is applied. "It is actually only the outside of the cement that gets contaminated."

These cement segments are now being tested under mechanical load after exposure to varying concentrations of hydrogen

⊞ **CIT Department of Manufacturing, Biomedical and Facilities Engineering:** http://www.cit.ie/Schools.cfm/type/page/ section/details/id/30/type/Page/action/pa ge/aID/191/CatName/Schools_&_Depart ments.html

A clue to defeating cancer cell growth

Galway research into the processes that allow a cell to make a perfect genetic copy of itself may aid the fight against cancer.

The two most important things a cell has to get right during its lifetime are the creation of daughter cells through division and the protection of the genetic code. Getting either of these crucial functions wrong can cause cell death, or worse still, lead to abnormal growth and cancer.

A research team at NUI Galway is delving into these two essential processes in a study based in the Department of Biochemistry. It involves a detailed analysis of what is going on during cell division or mitosis and when the cell carries out essential 'maintenance' on its DNA, says Dr Ciaran Morrison, who leads the research team.

"The cell has two important things to do when trying to divide," says Morrison, who received five-year Science Foundation Ireland funding worth almost €1.5 million to conduct the work. It must copy its DNA, and then accurately deliver identical copies to both daughter cells during mitosis.

"We are trying to use genetics and cell biology to investigate the links between the mechanisms that control maintenance of the chromosomes and the mechanisms that maintain mitosis," Morrison says. These two systems are co-ordinated but it remains unclear how they are linked and how—or if—they talk to one another.

Progress in the analysis depends on interfering with the normal cellular processes to see what effect this has on mitosis and DNA repair. This is done by disrupting individual genes to create mutations, and then assessing the impact on cell biology. "The bulk of our work is with genes that have a role in these processes," says Morrison.

Radiation and chemical substances are used to cause breaks and errors in the DNA, but the best way to introduce mutations is by direct intervention, targeting an individual gene, he says. It is challenging work but you know exactly which gene has been disabled and so will have a better understanding of what contribution the gene makes to essential cell processes.

"We work on tumour cell lines for all our work," Morrison explains. "Most of the really important genes are essential so there is not a great deal of difference between tumour and normal cells."

They use 'gene targeting' to tailor mutations into the gene of interest. This involves adding extra DNA, for example for antibiotic resistance, which is bracketed on either side by DNA segments that the cell can recognise. This 'plasmid' gets incorporated into the gene, but causes it to stop functioning, to mutate or to die. When a gene has been knocked out or mutated, the team can assess the impact of this loss on mitosis or DNA repair by direct observation. "We can see these things down a microscope," states Morrison.

"We can see the DNA repair factors and see how they respond to damage." In effect they are looking at the "real-time DNA repair capacity" of the cell. The research team wants to understand the contribution made by various genes associated with mitosis and DNA repair, the reason being that failures in these essential cell functions can often become a trigger for the development of cancers.

Aberrant genes can cause inappropriate cell division and growth, so this work should lead to a better understanding of how cancers might occur and progress, says Morrison. It should also point towards ways to increase the efficacy of chemotherapy treatments that damage cancer cells.

Morrison's research effort is also looking at telomeres, the structures found at the end of a chromosome. These have a role in maintaining DNA integrity and gradually shorten over a person's lifetime. Telomere elements are also being knocked out "to see if it affects repair or mitosis", says Morrison.

Thursday, 19 February 2004

NUI Galway Department of Biochemistry: http://www.nuigalway.ie/biochemistry/

Vaccine for virulent strain of hospital bug

A Trinity College research group has helped develop a vaccine against drug-resistant bacteria that can infect hospital patients.

A Dublin-based scientist has developed a working vaccine against a common bacterium that causes dangerous infections in hospital patients. He believes this is the first working vaccine against *Staphylococcus aureus*, the bug that has become resistant to some of our most powerful antibiotics.

"It is looking promising," says Prof. Tim Foster of the Microbiology Department in Trinity's Moyne Institute of Preventive Medicine. The vaccine has been used successfully in a Phase II medical trial of 500 low-birth-weight premature babies, reducing both mortality and infection rates, he says. "They are planning a Phase III trial later this year."

The vaccine grew out of research in the early 1990s when Foster and his group discovered a novel protein on the surface of the *S aureus* cell called the clumping factor. "A patent was filed because we realised the potential significance of this for developing a vaccine," he says. "At the time the research was funded by the Wellcome Trust." Later funding came from BioResearch Ireland, and more recently the Health Research Board, Science Foundation Ireland, Enterprise Ireland and the EU.

The novel protein meant that they had a target unique to *S aureus*, a bug that has caused death and illness in hospital patients around the world. These bugs are fellow-travellers with us, living harmlessly in the noses of half the population and also commonly colonising the skin. Some strains have become strongly resistant to the antibiotics designed to kill bacteria, however.

This means they can often survive frontline antibiotics, constantly challenging the drug companies to find new drugs that can kill these strains, known collectively as MRSA (Methicillin Resistant *Staphylococcus aureus*). The current antibiotic of last resort, vancomycin, still works against most MRSAs but the medical journals increasingly carry reports about *staph* resistance to this antibiotic. "There are a couple of new antibiotics," says Foster. "Resistance will develop very quickly, so we view vaccination as an alternative."

Foster began a collaboration with a company in Houston, Texas, called Inhibitex, a research-based firm that spun out of Texas A&M University. The company grew rapidly and is now based in Georgia in the US and Foster is on the company's scientific advisory board.

enrolling for Phase II. This vaccine is based on monoclonal antibodies recovered from engineered organisms.

The trial involving the premature babies was a 'proof of principle' for this passive

+ Professor Tim Foster with, from left: Deirdre Sharkey, Jennifer Mitchell and Dr Evelyn Walsh

Foster searched for other proteins that might serve as alternative targets for *S aureus* and another *staph* bacterium, *S epidermidis*. "We discovered a lot more of these proteins," he says, "cousins of clumping factor." The team noted sequences in the first protein and looked for related ones in the *staph* genome, mining it and quickly finding 10 more. "It is like an insurance policy in case anything goes wrong with the primary target, the clumping factor," Foster says. Together Foster, Inhibitex and other collaborators have filed patents on six further proteins.

Trials are under way on two vaccines, the one used with neonates and based on the use of purified antibodies recovered from human donors and another trial that has just cleared Phase I and is currently

transferred immunity approach. One in five premature babies weighing less than one kilogram at birth will get *staph* infections, says Foster, hence the group's interest in a vaccine. Yet any patient would benefit from a *staph* vaccine before heading into hospital for surgery or emergency treatments. Patients who have to use catheters and young people may be particularly vulnerable too. "They are all at risk of *staph* infection," says Foster. "Some 40 to 50 per cent of hospital acquired *staph* infections are caused by MRSA."

The great advantage is the vaccine delivers antibodies, so treatment can be given to patients who already have infections.

Thursday, 18 March 2004

+ **Moyne Institute of Preventive Medicine, TCD:** http://www.tcd.ie/Microbiology/index.html

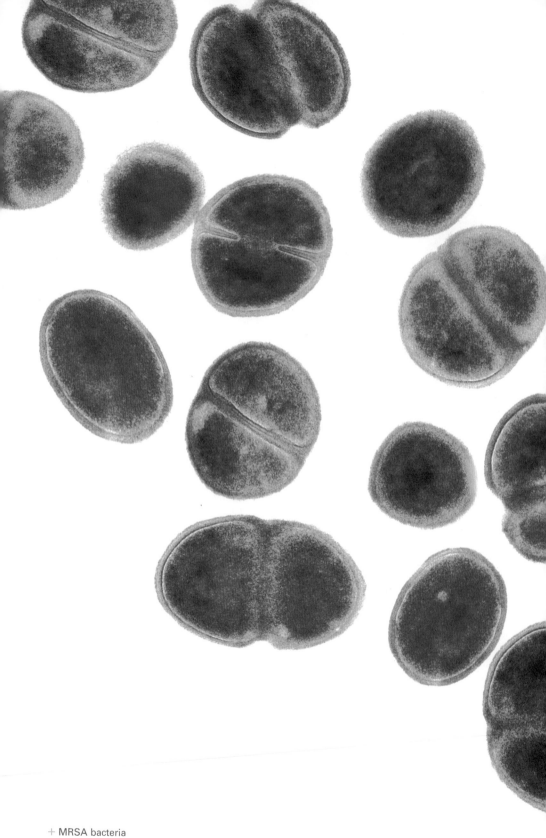

+ MRSA bacteria

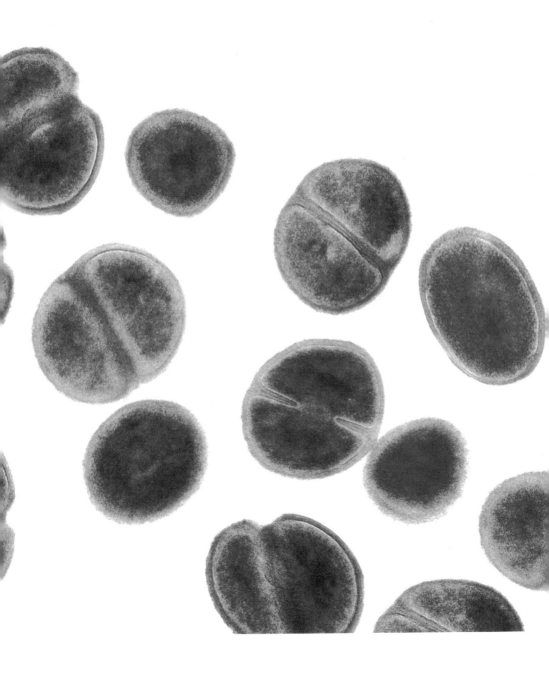

Science may change the way we eat

Algae, or at least parts of it, may soon be on the menu of those who aspire to a healthy diet. Algae on the dinner plate is just one of the changes that an ambitious new research programme hopes to introduce as a way to help reduce obesity in the EU.

Trinity College is co-ordinating a €16.5-million EU-wide research study of obesity and what can be done about it.

Research centres across Europe have banded together in a new effort to tackle the growing threat of obesity. The €16.5-million project has brought together a wide range of scientific disciplines, all focused on ways to stop or reduce the health impacts associated with our increasingly irresponsible dietary habits.

Researchers in Trinity College, Dublin have taken on the challenging role of co-ordinating this large-scale programme known as LIPGENE. It brings together researchers in 25 labs across Europe all working towards a common goal, to reduce the health risks associated with a number of obesity-related conditions known collectively as 'metabolic syndrome'.

"A factor of obesity is that people get a combination of a number of problems at the one time," says Trinity's Professor Michael Gibney in his explanation of metabolic syndrome. These include high blood pressure, changes in the amounts and types of fat circulating in the blood and conditions such as gout. Another key health impact, and one being faced by a rapidly growing number of people, is insulin resistance and Type II diabetes. "That is a big thing," says Prof. Gibney, given the alarming rise in incidence right across the western world.

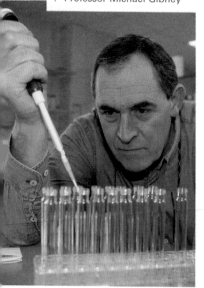

+ Professor Michael Gibney

Nutritionists and health specialists have discussed metabolic syndrome for decades but the term is gaining wider currency with the increase in those affected by it, an estimated 31 million EU citizens by 2010, says Prof. Gibney. "It has only really come out now with people recognising that obesity is a real problem."

Prof. Gibney and colleague Dr Helen Roche, both from the Nutrition Unit in the college's Department of Clinical Medicine, devised the work programme for the five-year LIPGENE project. They first proposed it five years ago but it was considered "too aggressive and too ambitious" under the then EU funding programmes. However, larger projects became possible under the EU's current research funding scheme, Framework Programme Six, and so LIPGENE was finally sanctioned, with the community providing €12.5 million and €4 million coming from other public and private partners.

LIPGENE is all about us, about how we live, the foods we eat and why we eat them. It is about why so many of the food choices we regularly make are unwise and it is about making it easier for us to eat foods that are better for us. "The whole philosophy behind it is if Europeans have a bad diet, there are two things we can do. We can change their diet or we can let them eat what they are eating now but make it better for them," explains Prof. Gibney. Getting people to make the right choices for themselves seems insurmountable, so LIPGENE is directed at the latter option.

"Technology can solve the problem," Prof. Gibney believes. "We are trying to find technological solutions which we can incorporate into the primary food production systems so people don't have to make changes themselves."

The algae research is a classic example of this approach. Most people accept that Omega-3 fish oils can help reduce the levels of saturated fats in the blood, bad cholesterols and other substances. Yet the fish themselves build up levels of these long-chain polyunsaturated fats (n-3PUFA) by eating algae. Plant biotechnologists working under the LIPGENE programme hope to isolate and clone the algal metabolic pathway responsible for these oils and then insert it in the linseed plant, delivering a fish oil supplement without the fish.

"From an ecological point of view this is terrific," says Prof. Gibney. Fish stocks are under threat and few people eat enough fish to benefit even if stocks were plentiful. With this approach consumers would have a good source of n-3PUFA without the environmental impacts associated with heavy fishing.

LIPGENE is also focusing in on our genes and their involvement in obesity, says Prof. Gibney. "We hope to be able to predict what individuals will develop metabolic syndrome if they follow a specific diet and have a given gene pattern," he says. Some individuals are highly sensitive to insulin resistance and this has been associated with a single gene. "That is only one gene," says Prof. Gibney. "We want to look at 150 genes and map them"—all of them expected to play some role in obesity and the development of metabolic syndrome.

Animal husbandry will also form part of the LIPGENE research programme, explains Prof. Gibney. The focus here is on dairy fats, a key cause of cholesterol-raising saturated fatty acids in the blood. Researchers will alter the bacterial populations in the stomachs of cows, helping to alter the types of fat formed in the milk. Dietary changes also offer a way to change fat levels in milk, allowing new kinds of milk and butter to be delivered to market. Alteration of the fat composition in poultry is another aim, with changes coming by feeding poultry with modified oils.

Another vital strand of the programme involves detailed social and economic studies of food consumption. Will people accept healthier food products if produced using genetic modification technology? Will they alter food purchases if they feel at risk of metabolic syndrome? Economists will also have to study the numbers associated with this whole area, says Prof. Gibney. They must study the costs to society of doing nothing, of introducing drug therapies to treat metabolic syndrome and of changing food production technology down on the farm.

Tuesday, 23 March 2004

➕ **Nutrition and Dietetic Studies Unit, TCD Department of Clinical Medicine:**
http://www.tcd.ie/Nutrition_and_Dietetics/index.html
The LIPGENE project: www.lipgene.tcd.ie

Telescope plan to reach for the stars

Ireland's position on the periphery of the continent could help radio astronomers right across Europe to see further into outer space.

A group of scientists have embarked on an unusual DIY project—drumming up the €10 million needed to build Ireland's first large-scale radio telescope. With its home planned to be the Birr Castle demesne, the instrument would allow Irish researchers to participate in important world-class astronomical research.

The ambitious scheme was launched in Dublin last week, appropriately enough with a fundraising reception at the Royal Irish Academy and later dinner. Dubbed A Radio Telescope for Ireland (ARTI) project, it seeks to bring local researchers into the 21st century of astronomical research using an instrument based at home rather than sharing facilities located abroad.

Surprisingly, ARTI's backers have not targeted the government as paymasters for the project. "We are going to pursue the ERDF (European Regional Development Fund), SFI (Science Foundation Ireland) and private and corporate donors," says ARTI project co-ordinator, Dr Denise Gabuzda.

"It is a really excellent project and has a lot to offer," she said enthusiastically before the launch. "This is an instrument that would be in demand right around the world."

An astrophysicist at University College Cork, Dr Gabuzda says the ARTI proposal involves raising funds to construct and operate a 32-metre radio telescope at Birr Castle on a site to be provided by the castle's owner, Lord Rosse. "The estimate is it could be built and outfitted for €10 million," she says. Ongoing maintenance and operation of the dish would then cost about €300,000 a year.

Siting the instrument at Birr has powerful historical resonance, given that it is the home of the spectacular 72-inch reflector telescope built and installed at the castle in the 1840s by the third Earl of Rosse, William Parsons. The wooden-tubed Leviathan was the largest telescope in the world for 75 years. The seventh and current Earl of Rosse has actively pursued this scientific heritage, fully restoring the Leviathan, opening a historic science centre and now offering a home for the proposed radio telescope. "Lord Rosse is interested in continuing the family tradition of supporting science and astronomical study here in Ireland," Dr Gabuzda says.

It might seem incongruous to link Ireland with astronomy given our persistent cloud cover, but radio astronomy offers a way around the weather. "It is an instrument we could do very useful work with even with the climate," she believes. These telescopes don't read light, they read radio signals given off by quasars, pulsars and galaxies. Interpretation of the radio signal as it varies over time gives information about the structure and behaviour of these distant bodies, and happily, the radio waves pass through our cloud cover without being impeded.

The proposed ARTI dish would be 32 metres across, not exceptional if working on its own but very powerful when paired with other radio telescopes across Europe. The Birr instrument would be used in conjunction with others, greatly multiplying its resolution.

The 32-metre dish would perform like a 3,200 km dish if paired with another that distance away. "At its highest resolution you could get 50 to 100 times the resolution of the Hubble space telescope when networked with others," says Gabuzda. For this reason other radio telescope sites in Europe would be particularly pleased if the Birr instrument was built, she says. Our position on the northwestern periphery of Europe greatly extends the distance between us and other European instruments, meaning better

resolution and the ability to see further out into the universe. "This is a very innovative project and would have a large impact on the radio astronomy that could be done in Europe," Gabuzda says.

+ This Cambridge telescope is the same size and general design as the proposed Birr telescope

She and project scientist Dr Aaron Golden of NUI Galway don't see this as the sole property of astronomers. "We would see this instrument used for educational purposes as well," she says. Second-level school students and university undergraduates should have access to encourage an interest in the sciences and astronomy in particular. The dish would also support tourism in the Birr area, and stimulate jobs in electronic engineering and communications technology, she believes.

Thursday, 1 April 2004

+ **UCC Astrophysics Research:** http://www.physics.ucc.ie/research.html#Astro
ARTI Project: http://www.physics.ucc.ie/ARTI.HTM
Birr Castle Great Telescope:
http://www.birrcastle.com/index.htm?mainFrame=http%3A//www.birrcastle.com/main.htm

+ Professor Lothar Steidler

Bacteria offer hope to Crohn's sufferers

A bacterium normally used by the food industry has found a new role in the treatment of debilitating human illnesses. After genetic modification the organism might provide a powerful treatment for inflammatory bowel disorders such as Crohn's Disease.

A UCC-based researcher has come up with a promising new treatment for inflammatory bowel disease based on the use of modified bacteria.

Human trials for the treatment are already under way at the Academic Medical Centre in Amsterdam. Earlier success in animal tests suggested "we have reasons to be hopeful", says Prof. Lothar Steidler of University College Cork, who pioneered the treatment. "It is a new pharmacological concept which can be compared to the development of the hypodermic syringe," says Steidler, a lead researcher in the Alimentary Pharmabiotic Centre and the BioSciences Institute at UCC. "The approach depends on the use of probiotic strains of bacteria from the gut. We are trying to use them as delivery vehicles for medication."

The work focuses on *Lactococcus lactis*, a bacterium used by the food industry, for example in the manufacture of cheddar. It is a well understood, safe, food-grade organism with no known pathology, making it ideal as a delivery system for drugs, says Steidler.

Funding for Steidler's potential new Crohn's treatment comes via Science Foundation Ireland's investigator grant programme. He moved across from Ghent University to work with Prof. Fergus Shanahan at the Alimentary Pharmabiotic Centre. The story actually began in Ghent, where Steidler worked in a specialist immunology lab that produced and characterised key immune system signalling proteins known as cytokines. "The original task was producing cytokines and developing ways to produce them," he says. "We came across *Lactococcus lactis* and decided to try and see what it could do."

The researchers engineered the organism, inserting an extra protein-producing gene, and it proved superb at being able to deliver high quality proteins. Steidler then had a 'eureka moment' when he realised that if engineered with the right cytokine the organism might provide a safe and effective treatment for inflammatory bowel disease (IBD). At about that time in the late 1990s, patients with Crohn's Disease were receiving treatments involving the cytokine interlukin-10 (Il-10). "Il-10 dampens down the inflammatory reaction," he explains.

"We knew it (the bacterium) was completely harmless, but on the other hand we knew it would make bioactive proteins," says Steidler. He decided to insert a cloned Il-10 gene into the organism, in the process knocking out the thyA gene, the thymidine-producing gene. Thymidine is an essential component of DNA construction and without it the bacterium simply self-destructs. "We established a system which would ensure us the strain was not viable outside the human or test animal," he says. This was essential given the widespread and "legitimate concerns" about any release of a modified organism, he adds.

They tested the engineered bacteria in a strain of mice that naturally has IBD. "We found we could cure the disease in these animals, it was quite striking," he says. "Next we wanted to try this in humans." The bacteria make an ideal delivery system for this disease. They survive the human stomach if encapsulated and pass through to the gut where they begin to produce inflammation-reducing Il-10 right where it is needed. They quickly run out of thymidine, however, and begin to die off, ensuring they can't survive and escape into the environment. "Everything we are giving is in complete containment," says Steidler. There are no natural repositories for thymidine in the environment so the organism can't survive outside.

Twelve patients are now receiving the modified organism as part of an early Phase I trial. They receive freeze-dried formations of *Lactococcus lactis* in capsule form daily, as the bacteria only remain viable for a short time in the gut where they deliver the Il-10. "We are only at the trial stage and I don't want to give people the impression that the cure has been found," Steidler adds, but the trial is running well.

He is already trying to improve the treatment, this time using a *Lactobacillus* strain discovered by UCC researchers. Even before modification it produces substances that help Crohn's patients and Steidler believes that if it also carried the Il-10 gene an even better treatment might arise.

Thursday, 08 April 2004

Alimenatry Pharmabiotic Centre, UCC: http://apc.ucc.ie/content/
BioSciences Institute, UCC: http://bsi.ucc.ie/
Academic Medical Centre, Amsterdam: http://www.amc.uva.nl/index.cfm?pid=1

Straight from the heart

Surgeons are getting very good at mending broken hearts by replacing faulty heart valves, but there is always room for improvement. An NUI Galway research team has helped with a new method for measuring blood coursing through artificial valves, providing information that could help prevent dangerous blood clots.

Researchers have developed a novel method of modelling blood flow through the heart, which should improve artificial heart valves.

The implantation of prosthetic or artificial heart valves is now commonplace, with about 275,000 patients worldwide receiving this surgical treatment each year. Yet the mechanical devices sometimes pose a risk of clots if blood cells become damaged when flowing through, says Dr Nathan Quinlan of NUI Galway. "One of the many engineering challenges prosthetic valves present is the difficulty of avoiding clot formation," explains Quinlan, a researcher in Galway's Department of Mechanical and Biomedical Engineering and the National Centre for Biomedical Engineering Science.

"The valves can give rise to unnaturally severe fluid dynamics in the blood that flows through them. This in turn aggravates blood cells and can trigger the coagulation process," he says. The usual medical response is to administer lifelong drug therapy to block clot formation, but Quinlan plans to use his new blood flow measuring technique to build a better replacement heart valve.

The method is known as "stereoscopic particle image velocimetry", he explains.

"It came out of the aerospace industry. It is one of a range of techniques to measure flow of a liquid or gas." It is no small feat to measure the way something like blood moves through the chambers of the heart. Cells can slow then accelerate as the valves open and close and the blood produces swirls and eddies that change in an instant.

These flows become increasingly complex when an artificial valve is inserted, something that also increases the 'shear' forces acting on the blood cells. If this reaches a point where cell damage occurs unwanted and unwelcome clots can form, says Quinlan.

The team, including Quinlan, Dr John Eaton and postgraduate student Mr Donal Taylor,

+ Dr Nathan Quinlan with postgraduate student Cathal Cunningham (left)

"We can see how the particles move from one image to the next. We get a measure of velocity in the fluid," he says. "Velocity in principle tells us everything. There are several things of interest, vortices, swirls in the flow and the shear stresses on the blood cells."

It sounds disarmingly simple but reading the data is anything but straightforward. "The size and density of the particles are important," he says, with the goal being to match particle density with fluid density. This keeps the particles in the flow and prevents them shooting off on a tangent as the flow turns and twists.

Typically, the overall flow is divided into layers moving at different speeds and exerting different shear stresses. The technique, which has been used to test valves supplied by industrial partners St Jude Medical in the US and Aortech in Scotland, can provide a detailed map of what is going on in the fluid. "This should aid understanding of complex fluid dynamics in the next generation of replacement heart valves," Quinlan says. It may also provide a new way to understand how blood flows through a damaged heart.

developed a way to use the aerospace technique in the biomedical area. It measures blood flow very accurately in a model heart fitted with various artificial valve designs. The Higher Education Authority via its Programme for Research in Third-Level Institutions provided funding for the research, which captured the bronze medal for the best bioengineering paper last January at the annual conference of the Royal Academy of Medicine in Ireland.

The method involves taking rapid-fire pictures of small particles suspended in the 'blood' flowing through the model heart. It uses a laser to illuminate the particles, taking 15 pairs of images every second as they move along in the flow, says Quinlan.

Thursday, 08 April 2004

+ **NUI Galway National Centre for Biomedical Engineering Science:** http://www.nuigalway.ie/ncbes/

+ An extract from *The Plough and the Stars* by Sean O'Casey

Fluther: Mollycewels an' atoms! D'ye
             think I'm going to listen to
             you thryin' to juggle Fluther's
             mind with complicated
             cunundhrums of mollycewels
             an' atoms?"

The Covey: There's nothin' complicated
             in it. There's no fear o' th'
             Church tellin' you that
             mollycewels is a stickin'
             together of millions of
             atoms o' sodium, carbon,
             potassium o' iodide, etcetera,
             that, accordin' to th' way
             they're mixed, make a flower, a
             fish, a star that you see shinin'
             in th'sky, or a man with a big
             brain like me, or a man with a
             little brain like you!"

Chinese snap up Irish design for Olympics

The physics of foam, Irish-style, has made its way into the spectacular Water Cube, national swimming centre for the 2008 Beijing Olympics.

What do the bubbles on a head of beer and the 2008 Beijing Olympics have in common? The Weaire–Phelan structure, of course, the most efficient construct yet devised.

Two Irish scientists currently hold the world record for the most efficient 'ideal' structure, one with the least possible wasted space between individual components. Now aspects of this design are to feature in the spectacular $100m (€86m) 'Water Cube' swimming centre to be built in Beijing.

"It was a discovery we made at the end of 1993," says Prof. Denis Weaire, head of the Physics Department at Trinity College Dublin. He and Robert Phelan, then in his first year as a research student, picked up a research challenge laid down more than a century earlier by none other than the great British scientist Lord Kelvin. The challenge is deceptively simple, says Weaire. "How can you divide up space into equal volumes all with a minimum surface area? Ideally that is what a foam should do in foam physics," he says. "This was a celebrated problem. Lord Kelvin posed it in 1887 and gave his own conjectured answer."

The rules hold that each bubble or structural component must have an equal volume. Yet what shape should they be to reduce to a minimum the wasted space around the edges and between bubbles? "What we produced on a computer was a new ideal structure for foam, a foam made up of identical bubbles," says Weaire. Their design beat Lord Kelvin's effort, a record that had stood for more than a century. "We were lucky enough to come up with a structure that had a lower surface area," says Weaire. "It caused quite a stir among mathematicians and those physicists interested in foams. In more than 10 years no one has come up with a variation that does better than ours."

In physics terms the idea is to achieve the minimum surface area, in effect achieving a minimum energy structure. The Weaire–Phelan structure is mainly made up of pentagons, with a smaller number of hexagons of equal volume thrown into the mix. "It turns out that mathematically the hexagons are necessary," Weaire explains. "It probably is the ideal structure but there is no mathematical proof of that."

Here enters the Chinese connection and the Water Cube. About a year ago a Chinese engineer contacted Weaire, asking him about the Weaire–Phelan structure. "He didn't tell me what it was for and I forgot all about it until I saw it on the Internet," says Weaire.

Consulting engineers Arup announced that it and partners, PTW Architects, the China State Construction and Engineering Corporation and the Shenzhen Design Institute, had won the international design competition for the Water Cube. The Arup site acknowledged that the Cube's winning design was based on aspects of the Weaire–Phelan structure.

Weaire always thought that their bubble design was "pleasing to the eye" and therefore of possible architectural merit. "At the time I did suggest it was interesting architecturally. I am delighted (this) has suddenly happened." He believes the Chinese saw two virtues in using this for their 17,000-seater Water Cube. They wanted a water theme and foam bubbles are mostly made of water. The structure itself was also attractive, making for a striking, eye-catching building.

The 70,000 square metre building plays on the Weaire–Phelan structure, built from three different steel nodes and steel members to be bolted together on site. These in turn will support the translucent ethylene tetrafluoroethylene (ETFE) skin of the building.

+ A model of the Weaire-Phelan equal-volume foam

Thursday, 29 April 2004

+ 'Water Cube' swimming centre, Beijing

[H$_2$O]^3

+ Shaun Mahony

Shortcut is found to locate hidden genes in DNA

A postgraduate researcher at NUI Galway is using artificial intelligence to find a genetic needle in a haystack.

A needle in a haystack just about sums up the ongoing search for unknown genes secreted along the huge strands of DNA being catalogued each day by labs around the world. An NUI Galway research student has found a short-cut, however, using computer-based artificial intelligence to dig out that hidden needle.

The goal is to be able to locate genes dotted along a DNA genome, explains Shaun Mahony, a second-year PhD candidate working in Galway's National Centre for Biomedical Engineering Science. "These sequences are millions and billions of letters long," he says. "Only 2 per cent of the human genome is part of a gene. It is very difficult to separate them out."

Late last month he won the inaugural Embark/Institution of Engineers in Ireland computer-science research achievement award. Co-sponsored by the IEI and Embark, operated by the Irish Research Council for Science, Engineering and Technology, the award aims to recognise top-quality work done by Embark scholarship researchers. About 90 Embark scholars were eligible for the competition and Mahony was one of 11 sent forward by each recipient third-level institution for a final public presentation of their research projects. Aside from the kudos of the award, Mahony also received a €3,000 bursary to further his research aims.

Mahony graduated two years ago with a degree in electronic engineering, but searching through biomolecules was already familiar territory. He began a bioinformatics project at the end of his BSc and expanded this into his current PhD effort. "Bioinformatics is the study of DNA sequences using computer programs," explains Mahony. "It is a very young field. These DNA sequences have only been around for the last three or four years."

The human genome is about three billion 'letters' or base-pairs long, and even a lowly bacterium has a genome about two million letters long, he says. The challenge is finding the important working DNA elements, the genes that produce the essential proteins used by the organism to sustain life. Bioinformatics is the tool being used in this worldwide effort and Mahony is playing his part, writing two computer programs that help dig out the genes from among all the so-called junk DNA. More importantly, he built-in artificial intelligence, enabling his two programs—RescueNet and Sombrero —to learn and improve their searching performance.

He explains how his programs work by drawing an analogy with ordinary language. A sentence in a book full of gibberish could be equated to a gene sitting in a very long strand of DNA. A sentence can be made up of many different words, but a few, for example 'a', 'the', 'is' and 'are', would very often be part of a typical English sentence. "It is the same in a DNA sequence," he says. "There are DNA words that are used more often in genes but not in non-gene sequences. My software picks out these words that are used more frequently."

RescueNet works on chunks two to five million base-pairs long. The software is first 'trained' by being given gene sequences that contain the tell-tale DNA words. It learns as it progresses, getting better and better at spotting where the genes lie in the genome. "My software is flexible enough to pick out the pattern of these words," says Mahony. It is about 90 to 92 per cent accurate, close to the current best systems that are 95 per cent accurate.

"My software has particular advantages in some genomes, the hard-case genomes," he adds. The DNA language is made up of four word elements referred to by the letters A, C, G and T. Mahony's hard cases include "high G-C genomes" that have a high proportion of G-C combinations. The Sombrero package is new and yet to be published in the scientific literature. It is designed to find 'gene switches' that initiate gene activity. "These switches or promoter sequences are very small compared to genes," says Mahony. Sombrero searches through DNA blocks about 10,000 base-pairs long for promoter sequences only six to 20 base-pairs long.

Both packages are based on software for artificial intelligence known as 'self-organising maps', a form of neural network algorithm, says Mahony. "It is good at clustering data, for finding regions of similarity," he adds. This also enables the software to learn as it goes along. "By the time it finishes training, it knows the (DNA) words it should search for."

Thursday, 13 May 2004

➕ **NUI Galway National Centre for Biomedical Engineering Science:**
http://www.nuigalway.ie/ncbes/

Code cracker triumphs in battle of wits

It is a battle of wits with huge rewards for the winner. Crooks are finding new ways to attack 'smart card' bank- and credit-card technology while card manufacturers counter with increasingly complex defences.

A computer scientist at DCU is working out how to make smart cards safer— with an unexpected spin-off.

A Dublin City University graduate student is engaged in this battle royal while working towards a PhD on computer security. Claire Whelan, of DCU's Computer Science Department, benefits from a research collaboration agreed between the university and Gemplus, a Luxembourg-based smart-card manufacturer. It is all about protecting the security systems built in to smart cards, explains Whelan. "These are plastic cards with an embedded microprocessor." Examples include the SIM card in your mobile phone, the next generation of credit cards and, in the future, passports and driving licences, she says.

The chips carry highly sensitive coded information of value to thieves, providing a financial incentive to overcome each new security barrier as it is installed. Whelan is working on both sides of this fence, trying to crack systems as a way to make them safer.

"We are finding new attacks and defending against them," she says. "There are counter-measures, but nothing can guard against them 100 per cent." Even so, Whelan must be a formidable investigator, given her recent involvement in a new method to read blacked-out sections of declassified government documents. She and Dr David Naccache of Gemplus decoded the hidden words in a memo to President Bush that was released last month for an inquiry in to September 11th. The pair also read concealed words in documents from the Hutton Inquiry into the death of the UK government scientist Dr David Kelly.

The two devised a way to measure the width of the blacked-out word, first identifying the font, then measuring the

Declassified and Approved
for Release, 10 April 2004

trike in US

dia reports indicate Bin Ladin
ist attacks in the US. Bin Ladin
and 1998 that his followers would
mber Ramzi Yousef and "bring

Afghanistan in 1998, Bin Ladin
in Washington, according to

n Ladin Determined

ndestine, foreign government,

emonstrate that he prepares
rred by setbacks. Bin Ladin
and Dar es Salaam as early
planning the bombings were

US citizens—have resided
up apparently maintains a
-Qa'ida members found guilty
Africa were US citizens, and a
90s.

n Ladin cell in New York
attacks.

the more sensational
service in
US aircraft to gain the
an and other US-held

continued

Declassified and Approved
for Release, 10 April 2004

+ Blacked-out sections of declassified
UK and US government documents

This work is some distance, however, from her real research interests, which relate to 'side channel attacks' on smart cards. The work focuses on the tiny electronic signals given off by a smart card while it is in operation. "When smart cards are put in to the reader they do certain operations," she says. "Information leaks naturally from the card. You can pick this up and find ways to take information from the card."

They use several techniques, including studying how much power the card uses. "We look down at the very low-level operations and try to make a correlation between the power used by the card and these low-level operations," she says. If they can make the connection they can work out the hidden encryption key used to conceal the data, which in turn will open up whatever details the card holds.

Whelan received a three-year, €60,000 scholarship to pursue the work from the Irish Research Council for Science, Engineering and Technology. "I wouldn't be here if it wasn't for them," she says.

Thursday, 27 May 2004

covered word down to a small fraction of a letter width. A computer interrogates an online dictionary to find words of similar width; then grammatical analysis rules out most of these. A simple human scan of the remaining candidate words will reveal the most likely hidden one. The context of the sentence helps this, she says. The two researchers presented their technique earlier this month at the Eurocrypt 2004 conference, in Switzerland, where they caused quite a stir.

+ **DCU School of Computing:**
http://www.dcu.ie/computing/index.shtml

+ Array of atmospheric Cerenkov imaging telescopes

Exploring the secrets of the Black Holes

New Earth-bound telescopes that will help us understand black holes are being built in Arizona with help from Irish research teams.

When it comes to signals from deep space they don't get much more powerful—or enigmatic—than the gamma rays that pepper our upper atmosphere. Now a €15-million collaboration involving scientists from the US, Ireland and Britain hopes to unlock the secrets to be found in gamma rays.

The project is known as VERITAS, for Very Energetic Radiation Imaging Telescope Array System, says Dr John Quinn, a lecturer in experimental physics and head of the very high energy astrophysics group at University College Dublin. The collaboration involves building not one, but four powerful optical reflector telescopes, each with a mirror measuring 12 metres across. The first prototype telescope is under construction and will be installed by this October, says Dr Quinn.

All four will sit in Horseshoe Canyon, 1,770 metres up along the side of Kitt's Peak in southern Arizona. They won't see any gamma rays, which are invisible, but will be able to watch the impact they have on Earth's thin atmosphere 10 kilometres up, information that will help scientists understand the rays and where they come from, says Dr Quinn. "The difficulty in doing gamma-ray astronomy is the gamma rays are absorbed by the atmosphere," he explains. They produce a short-lived cascade of light when they smack into the atmosphere, however.

They cause an 'optical disturbance' known as Cerenkov light, says Dr Quinn. The telescopes pick this up as a very faint but characteristic bluish glow that can be analysed to give information about the source of the rays. "We don't detect them directly, we detect them through the optical disturbance," he says. "The array gives us much better positional information and information about the energy of the gamma rays."

The real trick is being able to see and record the Cerenkov light, however. "The telescope must be very large to detect this light," he says. It is akin to trying to spot a weak camera flash from 10 kilometres away. They are also difficult to spot because the cascades happen so fast and come infrequently. A typical optical disturbance lasts only six billionths of a second and existing telescopes only pick up about four rays per minute coming from the Crab Nebula, a supernova remnant that emits gamma rays.

The energy level tells something about the source of the gamma rays, he says. "There

are a whole range of scientific targets," including "galactic nuclei"—super massive black holes sitting at the centre of many galaxies—supernova remnants and pulsars. "The mere detection of this alone is quite important," says Dr Quinn. "It teaches us about the physics of these objects. There is also a lot of indirect information we can learn from this."

The Republic of Ireland is contributing money and expertise to this effort, with financial support coming originally from Enterprise Ireland and more recently via the basic research grant programme now administered by Science Foundation Ireland. US funding is from the Department of Energy, the National Science Foundation and the Smithsonian Astrophysical Observatory. The collaboration includes the University of Arizona, Harvard and the University of Leeds. Here it includes UCD, NUI Galway and Galway-Mayo and Cork Institutes of Technology.

"The UCD group has developed the data acquisition software for the telescopes," says Dr Quinn. Galway has developed the mirror alignment system and Galway-Mayo is working on a specialised camera system for monitoring weather information during Cerenkov light observations. CIT is also contributing to weather information, developing a specialised pyrometer to measure infrared radiation in cloud cover. "The effect of weather conditions is very important in this," says Dr Quinn.

Thursday, 03 June 2004

UCD High Energy Astrophysics Group: http://ferdia.ucd.ie/home.htm
Find out about the **VERITAS project** at http://veritas.sao.arizona.edu

Protecting metal memory beneath the sea and the skin

A team of scientists and medical professionals is collaborating on research into new measures to combat the problem of corrosion.

Oil rigs and medical implants have much in common, at least for one researcher at NUI Galway's National Centre for Biomedical Engineering Science. The link between the two is corrosion and how to stop it.

Dr Liam Carroll, lecturer in chemistry and a research director at the centre, used to study ways of slowing the decay of offshore oil rigs, caused by chloride in sea water eating into the metal. Now he is trying to find ways to stop, or at least slow, the degradation of medical implants in the almost-as-salty environment of the human body.

The centre received a €32million award from the Programme for Research in Third-Level Institutions, administered by the Higher Education Authority. Dr Carroll heads the biomaterials group, which includes a team of 25 researchers. They are looking at all types of implants, he says. "That covers everything from polymers to metals that go inside the body. One of the areas we are working on is looking at stainless steel and a new metal, nitinol."

Nitinol, a 50/50 blend of nickel and titanium, is a valuable new material for use in implants. It is already used widely in spectacle frames because of its 'metal memory', the ability to take twisting and distortion but return to its original shape, explains Dr Carroll. This makes it ideal for use in stents, implants used in a range of medical procedures to support damaged blood vessels and to open up clogged arteries. "We are looking at stents made from stainless steel and nitinol and used in the body. All of these things are subject to attack from chloride ions," says Dr Carroll.

Medical implants must undergo rigorous testing to ensure they are safe to use, he says. This includes ongoing tests for corrosion damage and an assessment of any possible consequences for the patient. It isn't all about protecting the implant from damage, he says.

Nickel is a carcinogen and 25 per cent of the population has some level of allergic reaction to the metal. While the implant has to be protected from gradual corrosion, the patient also has to be protected from the unintentional release of nickel, particularly when using the nitinol alloy, which is half nickel.

"The test methods are exactly the same," whether working on oil platforms or implanted joints, says Dr Carroll. The goal is to isolate the metal, finding ways of keeping chloride ions away from the surface. On an oil rig or ship's hull you bolt on 'sacrificial anodes', metals such as copper or zinc that protect the steel but gradually dissolve away in the process. He was experimenting with other metals, including aluminium, to find alloys that could serve as a sacrificial anode but corrode more slowly than copper or zinc.

With implants it involves getting a very thin oxide layer onto the metal surface, says Dr Carroll. Surgical stainless steel has an oxide layer only three to five nanometres thick.

"As long as that layer is intact everything is all right," he says. He and his team are looking at ways of chemically enhancing the oxide layer in what is known as a 'passivation treatment'. While the approach can make the oxide layer more resistant to corrosion, the treatment also "tends to thin the oxide", says Carroll.

The Galway team researches novel ways to block corrosion but also characterises new materials being developed by companies that manufacture implants. The work has

+ Expertise acquired in the study of the corrosion of offshore oil rigs is now being applied to medical implants.

become more complicated in recent years given the increasing number of 'drug-eluting' stents coming onto the market. These implants carry a coating that gradually releases a drug into the tissues adjacent to the implant.

Thursday, 17 June 2004

⊞ **NUI Galway National Centre for Biomedical Engineering Science:**
http://www.nuigalway.ie/ncbes/

DCU research aims to end insulin injections

Cell transplants hold promise as a way to end the daily insulin injections needed to keep diabetics alive. Advanced research under way at Dublin City University aims to make this promise a reality.

A specialist team at Dublin City University is working towards devising a new, more effective form of transplanting insulin cells.

Dr Lorraine O'Driscoll leads DCU's Diabetes Research Group, based in the university's National Institute for Cellular Biotechnology (NICB). Her group is pursuing a range of research initiatives all directed towards the transplantation of insulin-producing cells as a way to eliminate or reduce the need for injected insulin.

"The only therapy for Type I diabetics is insulin injections. They save lives but have serious limitations," says O'Driscoll. One alternative therapy being used to a limited extent is full pancreas transplantation. The pancreas contains beta cells, the cells that produce insulin as a way to control glucose levels in the blood. "Full organ transplants are possible but the transplant surgeon at Beaumont, Mr David Hickey, says that of the 80 to 90 pancreata donated each year only about 10 are suitable for transplant," O'Driscoll explains. The others are 'wasted' because of a failure to get a good tissue match between donor and recipient.

O'Driscoll's work could end these difficulties. She is studying procedures used to isolate, preserve and then transplant not the whole organ but only the islets of Langerhans, clusters of pancreatic cells that hold the insulin producing beta cells. Islet transplants have been pioneered at the University of Miami, she says. She spent time in Florida studying the techniques developed there before returning to DCU. "We plan to establish the first human islet isolation unit in Ireland," she explains.

She now leads research that is progressing on a number of fronts, starting with the isolation of islets from donated pancreata. "The islets that contain the beta cells that make insulin only represent 2 to 3 per cent of the whole organ," she says. Only about 40 per cent of these islets can be recovered using the current best techniques, and O'Driscoll hopes to increase this yield,

+ Dr Lorraine O'Driscoll

because existing storage methods are inadequate, O'Driscoll says, something she hopes to correct.

Islet transplantation is the way to go to eliminate the need for insulin injections, she believes. Whole pancreas transplants require close tissue matching to prevent rejection. "The islet transplants don't require tissue typing, only a blood-group match." Islet transplants also only require minor surgery under a local anaesthetic while whole organ transplants represent major surgery. "The islets are injected into a vein and lodge in the liver where they produce insulin," she explains. The procedure is very new so only short-term results are available, but two-year results on 300 patients treated at the University of Miami show that 71 per cent of them remain insulin independent after islet transplantation.

Of the remainder, they are finding that the beta cells continue to produce a background level of insulin that can be supplemented by injection. "That low level of insulin is important to control our blood glucose levels," says O'Driscoll, as it helps to even out glucose highs and lows if the patient response to injected insulin is erratic.

The NICB, headed by director Prof. Martin Clynes, last month announced a €34-million investment programme, including a new purpose-built €10.4-million research centre. Due for completion in 2005, the centre will help the NICB to greatly broaden the scale of its research activities.

something that would help treat more patients. Her team is also looking at the progenitor cells that produce the beta cells. These stem cells exist inside the pancreas and 'differentiate' to become beta cells. How this happens remains unclear in humans but she has already had some success differentiating mouse stem cells that have gone on to produce insulin.

Cold storage of islets, or 'cryopreservation', is also under study at the NICB. Islets for transplantation must be used quickly

Thursday, 15 July 2004

⊞ **National Institute for Cellular Biotechnology, DCU:** http://www.nicb.dcu.ie/
NICB Cell Engineering and Tissue Differentiation Project: http://www.nicb.dcu.ie/celldiff.shtml

Duo put Radon test in the picture

An ingenious method of measuring radioactive gases on old mirrors or the glass of photograph frames has been developed by UCD scientists.

Irish researchers have developed a way to read the 'fossil record' left by radon, the radioactive gas. It provides a way to measure what radon levels might have been years ago, long before any modern measurements were taken.

The technique was developed and refined by Dr James McLaughlin, a senior lecturer in experimental physics at University College Dublin, and PhD student Kevin Kelleher in collaboration with Dr Christer Samuelsson of the University of Lund, in Sweden. McLaughlin, who is also head of the natural-radiation-studies group at UCD, is using the technique in a planned 'look back' at homes with high radon levels in co-operation with the Radiological Protection Institute of Ireland (RPII). He has already measured historic radon levels in a home in Castleisland, Co. Kerry, that had the highest radon levels recorded anywhere in Europe.

The ingenious method involves taking radiation measurements of glass surfaces found in the home, explains McLaughlin. Old mirrors or the glass in family pictures hanging on a wall are perfect targets for this measurement technique, which depends on recording polonium 210, he says. Radon is a naturally occurring radioactive gas that seeps up from the ground. Outside it dissipates harmlessly, but it can become trapped in pockets under floorboards or subfloors. The gas decays in the air, in turn producing daughter particles that, if inhaled, can deliver a radiation dose to sensitive lung tissues.

The production of these daughter particles is what provides the fossil record of radiation levels in these homes, says McLaughlin. "Radon in an enclosed space produces decay products in the air, and the first of these is polonium 128," he says. "If it lands on a surface, for example a mirror or picture frame, it has a very short half-life of just three minutes." It then breaks down to produce the next step in the decay path, bismuth 214, but in the process releases an alpha particle.

The alpha particle goes in one direction and the bismuth 214 in the other, says McLaughlin. "This recoil energy is enough

to implant it up to 50 or 60 nanometres (billionths of a metre) into the hard surface. They are deposited on all surfaces but in particular on hard surfaces." They in turn go on to decay into lead 210 and polonium 210, isotopes that can be measured in the glass, explains McLaughlin. "In particular we measure polonium 210, because it is an alpha emitter and is long lived."

The researchers use alpha-particle detectors very much like those used to detect radon in the first instance. They measure alphas given off by the polonium 210 in the glass surface, providing a historical record of radon levels in the room as they were years ago. Window glass can be measured, but sunlight and air currents near the window can affect levels. For this reason they prefer to test family pictures, which are easily dated and tend to remain in a fixed location over many years.

"The RPII has a large databank of radon measurements in Irish houses," he says. It recommends remedial action to reduce radon levels when they reach 200 becquerels per cubic metre of air (bq). The look-back programme involves requesting test access in any home that reached levels of 1,000 bq or more. "We can measure as far back as Victorian glass," says McLaughlin. Radiation emitted by radon daughters in glass older than this stabilises, ruling out measurements in older glass, he says.

New tests on glass from the Co. Kerry home that recorded radon levels of 49,000 bq showed that these high levels have persisted there for as long as 20 years, McLaughlin says. Other high radon homes show similar findings.

+ Dr James Mc Laughlin, right with Kevin Kelleher (research student)

Thursday, 22 July 2004

+ **UCD School of Physics:** http://www.ucd.ie/physics/index.html
University of Lund: http://www.lu.se/o.o.i.s/450
Radiological Protection Institute of Ireland: http://www.rpii.ie/

+ Pictured is the installation of the Millennium Spire on Dublin's O'Connell Street. Engineered by Cormac Deavy and the team at Arup Consulting Engineers (http://www.arup.ie/), the spire rises 124.8m above the streets of Dublin and weighs in at a massive 133.16 tonnes. The top 12-metre section of the spire is perforated with 12,000 holes, allowing for lighting.

Colonies of bacteria in war against pollution

Bacterial colonies could soon be pressed into service cleaning up pollution from the Irish food industry. If successful, the biological system could replace expensive chemical systems and in the process produce a useful fertiliser.

UCC and Queen's are collaborating on efforts to utilise bacteria to help clean up pollution from the Republic's food industry.

The work in University College Cork's Department of Microbiology is led by Dr Alan Dobson and involves a collaboration with colleagues in Queen's University Belfast. It is co-funded by the Environmental Protection Agency and the Department of Agriculture and Food. The goal is to adapt existing biological waste water clean-up methods for use in the Republic's dairy industry, explains Dr Dobson. In particular, the research group is targeting phosphates, nutrients that promote eutrophication of waterways and the runaway growth of algae.

"There are strict legislative controls on the amount of phosphates that can be released after processing," says Dobson. "Typically what companies and municipal authorities would do is add chemicals to precipitate out the phosphate." However, this greatly increases the amount of sludge left behind by the process, producing a new waste problem. Dr Dobson is looking at "biological phosphate removal", the use of colonies of bacteria that happily gobble up as much as 90 per cent of the phosphates in waste water.

"It is a widely accepted technology," he says, and is already in use in the US, Europe and South Africa. "What we are doing is seeing if this removal system can be applied in an Irish context to handle food-industry waste." Results so far look very promising, he believes. "We have small laboratory-scale bioreactors to mimic the sludge treatment systems."

The sludge is mainly made up of bacteria and the process involves switching from anaerobic to aerobic conditions to encourage phosphate uptake. If the bacteria are held in anaerobic conditions before switching to aerobic, they undergo what is called a 'luxury uptake' of phosphate, boosting absorption to

+ Dr Alan Dobson

says, probably because of the interaction between organisms in the sludge. He is using DNA-based techniques, including polymerase chain reaction (PCR) analysis, to help identify the organisms that live in the sludge. The researchers are also testing whether it will be possible to use the sludge as feedstock for compost. The organisms grab hold of large amounts of phosphates and if added to any natural compost should boost levels of this important plant nutrient. The team put together a lab system to monitor quality of compost made in this way to see if it is safe and can provide benefit.

The collaboration with Queen's arises from earlier work done in Belfast showing that phosphate-loving bacteria increase their intake when 'shocked' by more acidic conditions. The UCC team is trying to apply this to its phosphate uptake system to improve performance and is also helping Queen's to identify the bacteria in its own biological clean-up system.

Dr Dobson's work shows that such a treatment system could be retrofitted to existing systems, sitting close to the top of a wastewater stream and taking out the phosphates before later treatment processes.

Thursday, 19 August 2004

between 10 and 20 per cent of normal dry-cell weight. A large part of the work involves trying to identify the bacteria doing most of the clean-up work in the sludge. There is a complex mix of symbiotic organisms living in the sludge, which together make an effective phosphate removal system, and Dobson wants to know who lives in the neighbourhood. Unfortunately, the active organisms have proved notoriously difficult to culture, he

➕ **UCC Department of Microbiology:** http://www.ucc.ie/ucc/depts/microbio/

Queen's University Belfast Department of Microbiology and Immunobiology:
http://www.qub.ac.uk/cm/mi/

'Blushing' brain offers hope for paraplegics

A Maynooth scientist has discovered a way to help accident victims who lose control of their limbs.

Watching the brain 'blush' offers a promising new way for paraplegics to interact with the outside world. It enables the user to simply think about something as a way of answering yes or no or to trigger a switch. The system was developed at NUI Maynooth in a project headed by Dr Tomás Ward of the Department of Electronic Engineering. It involves collaboration with colleague Dr Charles Markham of Maynooth's Department of Computer Science and Dr Gary McDarby of Media Lab Europe, with funding coming from the Higher Education Authority. "We are interested in brain/computer interfaces," explains Dr Ward who is developing the project with graduate student Shirley Coyle.

Their research involves finding new ways to communicate for patients and accident victims who lose the use of their limbs. Much research work has gone into new communication systems based on reading brain waves generated by thought, but these methods have proved difficult to achieve, explains Dr Ward. The signals can be a problem to read and interpret and the systems are difficult for the patient to use.

"It takes the user months to learn how to use these systems. They take incredible mental energy to use," he says. "We wanted to come up with a simpler approach that worked." The researchers employed a "totally different modality" simply by shedding a little light on the problem. The light in this case comes in the form of near infrared wavelengths delivered by light-emitting diodes. The near infrared (NIR) light is akin to that emitted by your television remote control, but has the advantage of being able to penetrate skin and bone easily. It is used to illuminate brain tissue, which can then provide information about brain activity, Dr Ward explains.

The whole idea is based on the fact that oxygen demand goes up when different regions of the brain are activated to control thought or motion, says Dr Ward. Specific regions of the brain are associated with particular activities, whether it is moving your hand, counting or speaking. The NIR in effect illuminates blood cells

+ Dr Tomás Ward, left, and Dr Charles Markham
photographed with the world's first Optical Brain Computer

flowing in the brain, he says. The red blood cells that ferry oxygen throughout the body look slightly different under NIR light depending on whether they hold or have shed their oxygen supply.

Dr Ward's system detects these differences, enabling the researchers to watch brain activity in real time. "You can see as the brain is working it 'blushes' slightly," he explains. "The system sees the brain blushing in different regions as these regions become active." Coyle developed NIR sensors that rest on either side of the skull over the regions of the brain linked to hand movement. The sensors could detect brain activity on either side of the brain no matter whether the subject moved their hand or simply thought about doing so. This represents a convenient way to provide a yes/no or on/off signal with the subject doing no more than thinking of moving their right or left hand. "That gives you a nice little communications device," says Dr Ward. The work provided the first proof that a NIR light source could be used in this way to produce a readable signal of brain activity.

"The problem with the system is it is slow. It takes three to five seconds to get a response," he says. "The great thing though is the learning curve is basically zero." The patient can master use of the system within minutes. The researchers published their findings recently in the *Journal of Physiological Measurement*. "We have demonstrated that this in principle works," he says. It opens the way for a company to develop new products based on the technology developed at Maynooth.

Thursday, 26 August 2004

+ **NUI Maynooth Department of Electronic Engineering:** http://www.eeng.may.ie/
NUI Maynooth Department of Computer Science: http://www.cs.may.ie/

Shirley Coyle, Tomás **Ward,** Charles Markham and Gary McDarby, (August 2004) On the suitability of near-infrared (NIR) systems for next-generation brain–computer interfaces, *Journal of Physiological Measurement* 25(4), 815–22.

Journal of Physiological Measurement: http://www.iop.org/EJ/journal/PM

+ Dubliner Ronan O'Toole spent five weeks in the Earth's deep freezer in the company of penguins, seals and a handful of humans. The microbiologist went out to Cape Hallet, Antarctica, to study bacteria in the sea ice and came back committed to the preservation of this unique ecosystem.

Holy grail of artificial life

Researchers at Dublin City University are busy constructing a completely new life form, one based on computer software rather than cells. This 'living technology' will grow, reproduce and evolve just like the real thing, however, and in time may help to produce computers and robots that can repair themselves.

Researchers at DCU are trying to produce a type of 'artificial life' using software that can learn and evolve.

A form of "life, but not as we know it" is the simplest way to understand the work under way in DCU's Research Institute for Networks and Communications Engineering (RINCE). Dr Barry McMullin leads DCU's €8.5-million contribution to a wider EU research initiative called PACE, Programmable Artificial Cell Evolution. Its goal is to build artificial life forms. "It is taking an engineer's view of the problem of evolution," explains McMullin. "How do we build robust, self-replicating computer systems." The closest thing to this goal at the moment is that bane of computer-users everywhere, the software worm. These self-perpetuate and bore into computer files, damaging information as they go. The aim of PACE, funded by the EU's 6th Framework

+ Dr Barry McMullin

Programme, is to produce self-sustaining 'software agents' that can change and evolve as the computer environment in which they live changes, says McMullin. "They are self-sufficient little computer programs that can grow, replicate, diversify within the computer system," he explains. "The field is known as artificial life," and the "holy grail", he adds, would be the development of the "spontaneous, evolutionary growth of complexity in a computer system".

The engineers want to model such systems on biological life, which evolves as its environment changes. Some changes end a biological life form, others increase its chances of survival, but these changes also increase the level of complexity within it. A computer is more complex than a stone but a single bacterium is more complex than a computer, he says. "The biological process leads to a spontaneous growth in complexity. We would like to achieve something similar in computing."

The approach being taken does not rely on applying artificial intelligence to help a software agent to 'think' its way out of a problem in order to survive. Rather, like a biological agent that uses whatever resources happen to be available to it, the software agent spontaneously incorporates small programme elements into its overall makeup and sheds others in response to changes in its environment. "This is a fundamentally new way to engineer software systems, by growing them rather than designing them," says McMullin. "The most complex machines we have don't do this." One benefit would be a computer system or program that can repair itself. By extension this might lead to a robot that can correct its own errors or learn from its mistakes.

McMullin also takes a wider view, suggesting that if we could build machines that can evolve then we should gain new insights into evolution itself. "If you could synthesise living organisms then you would get a better understanding of how they are organised and how life originated."

PACE is a four-year programme involving 13 research partners and two co-operating groups from eight European countries. The consortium includes chemists who will help create the microscopic chemical elements that should allow the artificial cells to assemble themselves automatically from non-living material. DCU's involvement focuses on trying to do all this in software simulations, models that will help to explain what can be achieved in the long term.

Thursday, 02 September 2004

➕ **DCU Research Institute for Networks and Communications Engineering:**
http://www.rince.ie/
PACE Project: http://134.147.93.66/bmcmyp/Data/PACE/Public

Ireland 'never linked to Scotland'

Ireland was always an island and a land bridge never formed to connect it to Britain, according to new research from the University of Ulster. Contrary to the general view, sea levels never fell far enough to allow dry land to emerge between the two landmasses.

A land bridge never formed between Ireland and Scotland, according to controversial new research.

This controversial new theory is proposed by Dr Andrew Cooper of UU Coleraine's Centre for Coastal and Marine Research. Working in co-operation with US colleagues from the University of Maine, the team has found compelling evidence the bridge never formed. There is no doubt there was a land bridge between Britain and the Continent 10,000 years ago, at the end of the last Ice Age. There was also a universal assumption that a second land bridge connected Ireland to Britain, but Cooper's 'seismic stratography' data, which make it possible to visualise layers of sediments below the seafloor, suggest otherwise. This seismic work indicates that sea levels never dropped far enough to produce a land bridge to Scotland, says Cooper.

The last Ice Age tied up huge volumes of water in glaciers that covered the land, as a consequence dropping sea levels globally by as much as 130 metres, he explains. This should have been more than enough to connect us up to Britain and through it to continental Europe, but Cooper's work suggests the sea level fall around Ireland was only about 30 metres. The weight of ice here and in Scotland was enough to

depress both land masses, pushing them down and keeping us separated by water. When the ice retreated and sea levels rose, the land rebounded, maintaining the waterway between us.

"Our geophysical surveys suggest that 10,000 years ago the sea level drop was 30 metres and 30 metres isn't enough to form a land bridge between Ireland and Scotland," he says. "For a land bridge between Ireland and Scotland you would have to drop water levels by 60 metres." The two universities are hoping that seabed sediment cores retrieved during the summer will help prove their theory. Last June they used coring equipment brought in from the US and

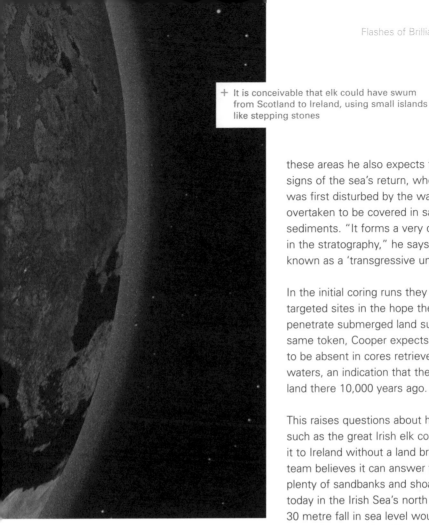

+ It is conceivable that elk could have swum from Scotland to Ireland, using small islands like stepping stones

these areas he also expects to see clear signs of the sea's return, when the land was first disturbed by the waves and then overtaken to be covered in sand and later sediments. "It forms a very distinctive layer in the stratography," he says, a feature known as a 'transgressive unconformity'.

In the initial coring runs they intentionally targeted sites in the hope they would penetrate submerged land surfaces. By the same token, Cooper expects these layers to be absent in cores retrieved from deeper waters, an indication that there was no dry land there 10,000 years ago.

This raises questions about how animals such as the great Irish elk could have made it to Ireland without a land bridge, but the team believes it can answer this. There are plenty of sandbanks and shoals known today in the Irish Sea's north channel, and a 30 metre fall in sea level would have been more than enough to lift many of them out of the water to become islands. Movement from island to island "would have made passage by sea much easier", he says. The elk could well have reached these islands and ultimately Ireland by swimming between them according to the University of Maine researchers. "They have observations of moose swimming out five or six miles to reach islands in large lakes," says Cooper. "It is conceivable that the elk could have done this across the islands, using them like stepping stones."

carried by the Northern Ireland Department of Agriculture's research vessel, *Lough Foyle* to retrieve a series of seabed core samples from Belfast Lough and from along the northern coast near Portrush, Co. Antrim. The team plans to open the first cores later this month.

"The submerged landscapes were dry land about 10,000 years ago, at about the same time as the first humans arrived in Ireland. People most likely lived and hunted on these ancient coastal plains," says Cooper. "As the sea level dropped, land plants colonised the land. If we are really lucky we will pick up some peat horizons or other organic horizons showing dry land." Over

Thursday, 09 September 2004

⊞ **UU Coleraine Centre for Coastal and Marine Research:**
http://www.science.ulster.ac.uk/ccmr/

Shining a light on the treatment of cancer

A University of Limerick graduate heads an EU research team based in Italy that uses light as a way to detect cancers. The same method is being applied to the detection of toxins and harmful chemicals.

An Irish scientist is head of a leading European Union laboratory in Italy where light is being used to diagnose cancer.

Prof. Maurice Whelan is based at the EU's Joint Research Centre in Ispra, Italy. There he leads the 'Photonics Sector' within the Institute for Health and Consumer Protection (IHCP). In effect, he uses light signals as a way to spot specific proteins that can indicate early cancers and detect changes in cells both *in vitro* and *in vivo*. He and other EU scientists are involved in a major community briefing on cancer research in Europe next Monday in Amsterdam. They will describe the work being done to improve the diagnosis and treatment of cancers.

Whelan's latest work is aimed at developing a high-throughput, automated method to test for chemicals and assess their toxicity to healthy tissues. "This robotic automated system is a step towards a miniaturised laboratory on a chip," says Whelan, the idea of acquiring a wealth of biological information from a single sample using a device no bigger than a microchip.

Whelan completed his PhD at UL in 1990 and took up post-doctoral research at Ispra. "I came here originally on a Marie Curie post-doctoral fellowship," he says. He retains strong links with the university, however, as an adjunct professor of biomedical materials in UL's Stokes Research Institute within the College of Engineering. He supports research opportunities at Ispra for UL PhD students, 15 of whom have spent part of their studies doing research at the IHCP. Now more researchers from UL will be travelling to Ispra after the university received Science Foundation Ireland funding for a project on early diagnosis of bowel cancers. "People in that project will be coming to us to focus on photonics," says Whelan.

Whelan is a specialist in photonics or optics. "We apply optical techniques for measuring and sensing," he explains. He was originally involved in laser metrology and later the development of sophisticated fibre optic sensors. One such project involved a whole-building sensor system that monitors subsidence and temperature throughout the cathedral in Como, northern Italy. Since then his research has moved towards nano-biotechnology, the development of biosensors and biomedical imaging systems that allow the analysis of tissues and cells, both *in vitro* and *in vivo*. One approach is based on the use of 'fluorescent spectroscopy', the study of the fluorescence given off by some substances when exposed to ultraviolet light.

"Most organic materials like tissues have biomolecules that, when they absorb UV light, they emit longer wavelength light," he explains. "We use UV lasers and analyse

the spectrum of the light that is emitted." The technique is becoming increasingly sensitive and can identify the presence of individual proteins, provided they fluoresce. One case where this has proved particularly successful relates to a protein called PP-IX that accumulates in cells affected by inflammation, often a precursor to diseases such as cancer.

PP-IX fluoresces very well, and using this Whelan and colleagues at two research centres in Strasbourg developed a new endoscope, a powerful tool of value to surgeons attempting to visualise diseased tissues when treating pancreatic cancer. It allows them to identify and remove cancerous tissue while leaving healthy tissue behind. The device will soon undergo human trials at the Academic Medical Centre in Amsterdam.

Whelan has now been appointed project leader of an IHCP initiative to develop advanced techniques for chemical testing. It arises because of two EU directives, one blocking the use of animal testing by cosmetic manufacturers and another called REACH, better known as the chemical directive. His photonic approach will allow *in vitro* tests using cell cultures, monitoring changes to cell activity or shape after the introduction of chemical substances. The directive involves about 30,000 different chemicals and Whelan's goal is to develop a high-speed robotic toxicity testing system that can carry out the thousands of tests needed under these two directives.

Thursday, 30 September 2004

✚ **University of Limerick Stokes Research Institute:** http://www.stokes.ie/

EU Commission Joint Research Centre: http://www.jrc.cec.eu.int/

Institute for Health and Consumer Protection: http://ihcp.jrc.cec.eu.int/

Does sea spray keep earth cool?

The cloud cover and haze caused by minute bubbles bursting on the surface of the sea may help to slow the pace of global warming.

Bursting bubbles on the sea could serve as a brake on climate change. Researchers in Ireland and Italy have identified a previously unknown source for cloud cover and haze that can reflect back sunlight and so reduce global warming.

The bubbles themselves don't make the difference; it is the organic matter they inject into the atmosphere when they burst that is important, says physics lecturer Prof. Colin O'Dowd, of NUI Galway's Environmental Change Institute. The minute particles, known as aerosols, drift on the air and act like seeds for water droplets that accumulate to form clouds and haze. O'Dowd and Dr Maria Cristina Facchini, of the Italian National Research Council's Institute of Atmospheric Sciences and Climate, identified an unknown link between the growth of sea plankton and climate. They published their findings last week in the journal, *Nature*.

"What we found was that the largest source of aerosols in the marine environment is organic matter from the plankton, produced by bubble-bursting mechanisms at the surface," O'Dowd says. They were studying the contribution made by aerosols to climate and the formation of clouds. They recorded the aerosols coming on to the coast at NUI Galway's Mace Head observatory, an important sampling point for European environmental scientists.

The long-held assumption was that iodine- and sulphur-based gases released by organisms such as plankton at the sea's surface made up most of the aerosols released into the atmosphere by the marine environment, he says. Their findings turned this assumption on its head. The plankton's contribution is greatest in the spring and summer when warmer seas boost its growth. "This organic factor in the aerosol content has a large seasonal component," says O'Dowd. It never entirely goes away, however. Organic matter at the surface is flung into the atmosphere to form aerosols when small bubbles burst at the surface in whitecaps tossed up by the wind. The aerosols, another name for particles small and light

+ Professor Colin O'Dowd

enough to drift on the air, in turn become nucleation points for the water droplets that congregate to form clouds.

The finding is exceptionally important because the organic material represents a previously unknown source of aerosols. Its contribution is therefore left out of current models that attempt to predict future climate, says O'Dowd. "Climate prediction models are huge for the development of policies and strategies against global warming," he says. "They are well-developed in terms of atmospheric gases but they are very poorly developed in terms of aerosols and associated cloud impacts. None of the current models take this into account."

The research showed that there was a ten-fold increase in organic matter in the aerosols between winter and spring. Simulations run by the research team also showed that the extra organic-derived aerosols could double the cloud droplet concentration and represented "an important component of the aerosol/cloud/climate feedback system." This is why the

discovery should have a major impact on climate models. Current thinking on global warming suggests that warmer sea waters will encourage plankton growth. On the basis of O'Dowd's and Facchini's work, more plankton growth would mean more cloud cover and a greater potential for turning back solar radiation.

O'Dowd urges caution in relation to this feedback mechanism, however. There is no measure as yet of the strength of this feedback. Also, greenhouse gases live far longer in the atmosphere than aerosols, he warns. "There is the potential for some degree of feedback," he says. "This should not be taken as a licence to emit pollutants as we wish."

The research does, however, clearly demonstrate the link between the marine biosphere and climate change, and the way in which living things can have a positive impact on our warming planet.

Thursday, 14 October 2004

⊞ **NUI Galway Environmental Change Institute:** http://www.nuigalway.ie/eci/

Italian National Research Council Institute of Atmospheric Sciences and Climate: http://www.cnr.it/istituti/DatiGenerali_eng.html?cds=075

Colin D. **O'Dowd,** Maria Cristina Facchini, Fabrizia Cavalli, Darius Ceburnis, Mihaela Mircea, Stefano Decesari, Sandro Fuzzi, Young Jun Yoon and Jean-Philippe Putaud, (October 2004) Biogenically driven organic contribution to marine aerosol, *Nature* 431, 676–80.

Nature: http://www.nature.com/nature/journal/v431/n7009/abs/nature02959.html

+ " The largest source of aerosols in the marine environment is organic matter from plankton, produced by bubble-bursting mechanisms at the surface." Professor Colin O'Dowd

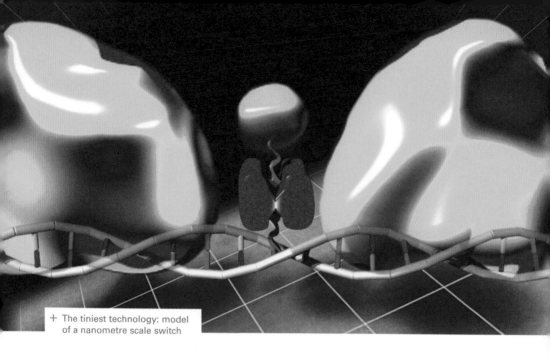

+ The tiniest technology: model
of a nanometre scale switch

At the shrinking
edge of research

An Irish research team is 'growing' microscopic wires
and switches that in time could find their way into the
next generation of miniaturised computers. The delicate
process involves the use of DNA strands, the building
blocks of all life on this planet.

Scientists at University
College Dublin are
utilising DNA technology
in order to grow
infinitesimal wires for
future computers.

Prof. Donald Fitzmaurice heads the work as director of
the nanochemistry group at University College Dublin.
His team is working in the nanometre range where
things are measured down to billionths of a metre or
millionths of a millimetre. These dimensions are almost
impossible to comprehend, Fitzmaurice admits. A single
human hair, at about half a millimetre across, is a
mammoth structure by comparison.

The research is all about finding ways to produce in a
predictable way tiny wires and functional switches and
transistors, essential components in any computer.
Electronic devices this small cannot be manufactured,
they are better grown in solution, Fitzmaurice explains.
He and his team have developed a process that can
reliably produce these components on a standard silicon
chip template. "We are getting something like 60 to 70

per cent yields on the template, which is higher than we thought," says Fitzmaurice. "It shows it is possible to build nanostructures with great precision that are important to the design of computer circuits using biological processes."

Their discovery has landed the group the cover story in the current edition of *Advanced Materials*, a leading international journal reporting on new materials. The cover shows the tiny junction point between two gold wires with a spherical particle of gold sitting in the gap between them. "This is a nanometre scale switch," explains Fitzmaurice. "It is a very small switch and has been built using the simplest methodologies." Small is too big a word for the device pictured here. It is measured in nanometres (nm), billionths of a metre.

The two wires produced by Fitzmaurice's group measure about 40nm long and 20nm wide. The minute gap between their ends is about 15nm across. The gold particle that sits fixed in this gap is about 10nm across, leaving spaces between the wire and the particle of between two and three nm. The gaps are important, says Fitzmaurice because they allow the nanostructure to function like a conventional transistor. These are the electronic components built into microchips to power computer systems. What is most striking about the research is that despite its precision, the nanostructure is made by dipping a piece of DNA attached to a silicon wafer into a series of solutions. "We have done all of that with beaker chemistry," says Fitzmaurice.

Fitzmaurice uses biochemicals as a way to dictate the nanostructure. Without these biological molecules the structure wouldn't

form. It begins with a strand of DNA, the genetic material from which all life on Earth is built, Fitzmaurice explains. DNA sequencers are now commonplace in modern labs so any DNA sequence required can be produced automatically to serve as a template. "We take a piece of DNA which is 90nm long that has a molecule called biotin attached to its middle," says Fitzmaurice. Biotin is a type of vitamin that in turn binds strongly to a protein called streptavidin.

The UCD team developed tiny gold particles that recognise DNA and can line up along it. When the silicon carrying the DNA strand is dipped in the first beaker containing the gold particles they form a gold chain 90nm long and 10nm across. "We then expose that gold particle chain to streptavidin," says Fitzmaurice. It binds so strongly to the biotin that it actually shoves some gold particles in the chain out of the way to produce a gap about 10nm to 15nm across. At this stage there are two gold chains lined up end to end with a small gap between. The next step is a form of electroplating that deposits extra gold along the chain, fusing the individual particles together to form two solid wires end to end with the gap still in place.

The final stage involves dipping this structure into a beaker containing individual gold particles each of which carries a small amount of biotin. This biotin fuses strongly to the streptavidin in the gap, in turn fixing the gold particle into position between the two wire ends. The electronic switch is complete, says Fitzmaurice. "This is a classic motif in the creation of nanostructures."

Thursday, 21 October 2004

⊞ **UCD School of Chemistry and Chemical Biology:** http://chemistry.ucd.ie/facilities.html

A. Ongaro, F. Griffin, L. Nagle, D. Iacopino. R. Eritja and D. **Fitzmaurice**, (October 2004) DNA-templated assembly of a protein-functionalized nanogap electrode, *Advanced Materials* 16(20), 1799–1803.

Advanced Materials: http://www3.interscience.wiley.com/cgi-bin/abstract/109712231/ABSTRACT

New discoveries in battle against arthritis

The mystery surrounding arthritis is gradually being unravelled and with a deeper understanding of the disease comes the potential for new therapies. Importantly, many new discoveries relating to the disease are being made here in the Republic.

Researchers are getting closer to what initiates the disease process in arthritis and to developing new drug treatments.

One of the country's top arthritis researchers provided a 20-year look-back at the disease in a public lecture that described the work being done here and elsewhere to understand what triggers this debilitating affliction. Prof. Luke O'Neill, head of the Department of Biochemistry at Trinity College Dublin, delivered an engaging talk on the subject last Thursday to a large audience at the Royal Dublin Society's concert hall in Ballsbridge. O'Neill studies the biochemical triggers that produce the joint inflammation seen with arthritis. Appropriately, the talk carried the title, *'Arthritis, how close are we to a cure,'* an approach that allowed him to track the progress being made in understanding its causes and in the development of new drugs against it.

The talk was organised jointly by *The Irish Times* and the RDS in association with the Irish Society of Immunology (ISI). Each year the Society honours one of its best researchers with an award given in recognition of work that contributes to a better understanding of immunology. Prof. O'Neill is this year's ISI award winner and so was invited to deliver the lecture.

"I have worked on arthritis for the past 20 years or so at various levels," Prof. O'Neill says. He worked with patients as a young researcher and in more recent years in the lab focusing on the disease in which joints become inflamed to produce pain. In some cases the arthritis is provoked by the body's own immune system. "That is what happens in arthritis, you get a deregulation in the inflammation process caused by the immune system," Prof. O'Neill explains. "The immune system is central to the development of arthritis."

There are two forms of the disease, rheumatoid arthritis and osteo-arthritis, with the former involving aspects of the immune system. Both, however, are characterised by inflammation in the joints. "In rheumatoid arthritis the joint is chock-full of immune cells," says Prof. O'Neill. His goal is to understand "what is going wrong in the immune system with arthritis," what is triggering the inappropriate immune-cell invasion and the release of inflammatory substances that follow it.

"What we do know now is we know the (immune) cells involved in joint inflammation and the chemicals these cells are making," he explains. "People are designing drugs to target these chemicals."

Prof. O'Neill's own research has continued along a single route from the beginning. "My goal has always been to get to the

+ Professor Luke O'Neill

"It could be that all rheumatoid arthritis is triggered by an infectious agent, but we haven't found it yet," he says. Infection by the bacterium that causes Lyme disease is known to cause joint inflammation. The problem is that if an infection actually is the trigger the agent that caused it is usually long gone by the time that joint inflammation sets in. "That is why people have had so much trouble tracking down the infectious agent," explains Prof. O'Neill.

Whatever the trigger, the result is that the immune system "begins to see the body's own tissues as foreign," he says. It launches attacks against certain tissues, in turn causing the joint inflammation seen in arthritis.

start of the arthritis process." The work has taken him step by step further up the cascade of events that lead to an arthritic outcome. "We now know what is driving the cytokines, they are known as toll-like receptors. We are getting closer to what initiates the disease process," he says. The work has also encouraged him to establish a campus company, Opsana Therapeutics, which will allow him to commercialise discoveries being made at Trinity College. "Academic research ultimately should be translated into the commercial world," he says.

Thursday, 28 October 2004

➕ **TCD School of Biochemistry and Immunology:** http://www.tcd.ie/Biochemistry/

Diving for buried treasure

The coastal waters bristle with all sorts of shipwrecks, from sailing ships to steamers, U-boats to Armada galleons. Now researchers from the University of Ulster, Coleraine, have catalogued these wrecks, detailing 13,000 of them in a book.

Deep-sea diving for shipwrecks on the ocean bed is a childhood fantasy for many. For two University of Ulster scientists it became an exciting reality.

Boats and Shipwrecks of Ireland is the work of Colin Breen and Wes Forsythe of the university's Centre for Maritime Archaeology. It is a comprehensive study of shipwrecks off Ireland's coastline, yet despite the extensive listing of 13,000 wrecks, there are thousands more waiting to be found. "We probably have 40,000 of these sites around our coasts," Breen says. "During the 19th century there were probably about 150 shipwrecks a year, which is a phenomenal amount."

All wrecks are of interest to Breen and Forsythe. "We ask ourselves, what is a wreck? The earliest boats we have are probably dug-out canoes. There are very different kinds of shipwrecks around the coast," says Breen. All are of value to marine archaeologists, he says, mentioning a 5,000-year-old Bronze Age wreck off the coast of Cornwall as an example. The vessel is completely gone now, but it was carrying a cross-channel cargo of bronze axes, which provide an 'artefact scatter' and evidence of the vessel's loss.

Wrecks can tell us much about marine technology at the time a vessel was lost, along with information about social structures, medicine, diet and much more, he says. A ship's captain and officers would have had the trappings of their class—books, crystal and good tableware—which would have contrasted with the possessions of an ordinary seaman. With a typical medieval wreck, the ship would have gone down carrying a three-month supply of munitions, food and just about anything else needed for a voyage. "It carried absolutely everything on board it needed to maintain itself," Breen explains. "Everything is condensed onto a single ship. It is like a floating town. You get this unique insight into the medieval period. It is like discovering Pompeii. The potential for

+ An abandoned seine net fishing boat on Dursey Island, Co. Cork

When asked about shipwrecks off Ireland most people immediately think about the ships of the Spanish Armada, 26 of which foundered off our northern and western coasts. Only seven of these have been found, says Breen and only three have been investigated. "They are extremely hard to find even with the use of modern technology," he says.

Breen dives on shipwrecks and describes the thrill of finding an underwater wreck. "It is that mix of childhood excitement mixed with academic discipline. You can get an incredible buzz from it." It is also highly dangerous, so the buzz is short-lived given the hazards of deep-water diving. Easy access to sites also means that they need protection from unauthorised disturbance by treasure hunters. UN agreements mean that countries must catalogue and protect maritime archaeological sites, he says. "Governments were essentially compelled to study in a systematic way the shipwrecks in their waters. There is an international drive to protect these sites," says Breen.

Thursday, 02 December 2004

maritime archaeology is unique for getting a momentary glimpse into the past."

The book carries details of the recorded wrecks, most of which date from the 1700s to the present. The oldest known European wreck dates back 11,000 years and the oldest here is about 6,500 years old. The book brings us right up to modern times with the listing of the *Lusitania*, among others. "The *Lusitania* wreck itself would not be of structural importance but it is of historical importance," says Breen. The more than 50 German U-boats sunk off Malin Head as part of Operation Deadlight in 1945 to 1946 are also featured in the book. "If we had just one on shore it would be a huge tourist attraction."

 UU Coleraine Centre for Maritime Archaeology:
http://www.science.ulster.ac.uk/cma/
Colin Breen & Wes Forsythe, (July 2004) *Boats and Shipwrecks of Ireland: An Archaeology*, Tempus Publishing.

Once we basked under the equator

Azure skies, bright sun, balmy temperatures and warm water lapping at your toes. The south of France? Majorca? How about Co. Cork 360 million years ago?

A study of rocks in Co. Cork reveals that millions of years ago Ireland had a Bahamas-like climate.

Stones don't say much but they do nonetheless have stories to tell. They can provide a great deal of information about past environmental conditions. A highly detailed study of bedrock outcrops across large swaths of Co. Cork reveal the region enjoyed a wonderful climate and once lay south of the Equator. It also shows that the rocks we see today spent millions of years at the bottom of the ocean.

Dr Ivor MacCarthy, senior lecturer in the Department of Geology at University College Cork, has published a 30-year study of the county's rocks in map form. The huge undertaking has required decades of work, given that it involves a detailed examination of rock outcrops across a 3,500 square kilometre area from Kenmare in Co. Kerry in the west to Macroom towards the east and all points south of there. The study began back in 1974, says MacCarthy. "It is a lot of work done over a long period of time. I have had quite a number of research students working with me over time," he says. "It gives detailed information on the distribution of different rock types across the region on a detailed scale. It gives very factual information about the rock types, but also the physical features of the rock, which are important in engineering terms," he adds.

+ Dr Ivor MacCarthy

While most people might be interested in what the rocks say about past climate, engineers will find the study invaluable because of what it tells them about rock conditions underfoot today. The area has many fractures and cracks that criss-cross the rock surface making it unstable, says MacCarthy. "They have quite a bearing on ground water and stability. The area is very heavily fractured. That gives the rock porosity and allows water to flow through but it also weakens the rock." His study has helped identify where rock instability is likely to occur in regions of bedrock hidden under the soil. The fractures can be up to 10 metres across and these gaps are filled with a variety of soil deposits. With flooding an increasingly likely prospect, given climate change, it is important for engineers, hydrologists and planners to know areas where flood induced ground collapse and rock instability are likely to occur, hence the value of MacCarthy's map.

The study, *The Geology of the Devonian-Carboniferous South Munster Basin, Ireland,* includes both a map and associated booklet. This provides explanatory notes and a detailed reference list of the findings, MacCarthy says. "The bulk of the rocks are sandstones and mudstones and some limestone." The rocks were laid down as layer after layer of sediments at a time when the region was between five and 10 degrees of latitude south of the Equator, he says. "The area was a little like the Bahamas."

The terrain was low-lying and so a shallow basin formed into which sediments were periodically washed. The conditions were generally arid during the Devonian period with infrequent, but occasionally heavy, rains and flooding. These filled the basin and it would then dry out. "These rocks range in age from about 360 million years down to about 350 million years old," MacCarthy says. "This is a really short period of time in terms of the overall age of the Earth."

Things changed at the start of the Carboniferous period, however, with a changing environment causing a rapid rise in sea level. The then freshwater basin that is now south Cork was flooded out by the sea to produce marine conditions. "Eventually the depth of water in the basin rose

dramatically to produce a relatively deep basin hundreds of metres deep," he says. "These major environmental changes are recorded in the rocks." Also there for those who can read them are the signs of what followed at the end of the Carboniferous 290 million years ago. The Earth's ever-restless crust began shifting the slab that now carries Africa northwards. This in turn crushed into the part of the crust carrying the rocks that now cover Cork.

The collision of the two great slabs threw up a colossal mountain range as high as today's Alps. North America was connected to Europe then and the range started at the US Appalachian Mountains, ran through Newfoundland and on across Ireland into southern Germany, says MacCarthy. The collision lifted Cork's rock cover out of the sea and also caused many of the fractures we see today, he says. Further cracking and splitting were caused by the 20 or so glaciations that occurred over the region since the Carboniferous. These glaciers also deposited soil and rock that filled the cracks and left the topography we see today.

The sedimentary rocks are about 8.5 kilometres deep under south Co. Cork, says MacCarthy. Sedimentary deposition continues apace in the North Sea, another sedimentary basin with rocks 10km deep. What is remarkable is that MacCarthy's study can interpret the clues that tell us the history of the stones we tread today.

Copies of the map and report are available for €25 by e-mailing i.maccarthy@ucc.ie.

Thursday, 06 January 2005

UCC Department of Geology: http://www.ucc.ie/academic/geology/

Policing proteins around the cells

Studying the 'motor cops' that carry nutrients and proteins into and out of the body's cells helps us understand how several diseases are caused.

Getting biochemical traffic into and out of cells is a complex business that is fundamental to health. When things go wrong with the mechanisms used to transport materials across the cell wall, diseases as varied as diabetes, BSE and cancer can be the result.

University College Cork cell biologist Dr Mary McCaffrey studies the 'motor cops' that regulate this essential traffic. She is a senior lecturer in the Biochemistry Department of UCC's Biosciences Institute and also a Science Foundation Ireland investigator with a budget of more than €1 million to conduct research into this area.

It is all about "how materials get through a cell membrane," Dr McCaffrey explains. Nutrients, iron, growth factors and signalling proteins move into and out of cells all the time in a tightly regulated system called endocytosis. "There is a very sophisticated system in place in order to move materials into cells. We work on the proteins that control the movement of these materials and also how they recycle," she says. "Recycling is a very specific term in cell transport. A protein moves from the cell surface into the cell where a decision is made whether to degrade the protein or recycle it back to the surface."

Dr McCaffrey likens each cell to an individual country with border police guarding the frontiers. Specialised transporter proteins sit ready to grab hold of materials and then guide them across the cell membrane and into the cell. Once they have the intended effect they are either recycled back outside the cell for reuse or broken down and discarded. Dr McCaffrey's lab is involved in the search for the transporter proteins and any other proteins that respond to them. She has discovered families of proteins known to oversee the process, including some of the group known as Rab proteins. Among others, she has studied Rab4, Rab25 and Rab11, with the latter two being particularly important given their linkages to cancers.

+ Dr Mary McCaffrey

Rab is known to be upregulated in many cancers and also affects the aggressiveness of the disease in the breast and ovaries, McCaffrey says. "The Rab protein dictates the aggressiveness of cancers," she adds. "The functional role of Rab11 isoforms in cancer is an area we are well placed to investigate with the extensive panel of reagents that we have developed over the years."

Rab is important in other diseases as well, she says. "Many diseases come about through a malfunction of membrane trafficking." She cites diabetes as one example, because Rab11 mediates the take-up of sugars into the cells. BSE is another, given one theory yet to be proven that the normal prion is converted into the abnormal, BSE-causing form after transport across the cell membrane.

She is currently seeking additional funding for the purchase of specialised equipment that will aid in her search for more Rab family members and the proteins they affect. One device is a form of 'molecular microscope' that will allow the team to monitor individual protein behaviour. "The more you understand the movement of materials within the cells the more chance you can control that movement," she says.

Thursday, 07 April 2005

+ **UCC Biosciences Institute:** http://bsi.ucc.ie/

+ Airfield Gardens, Dublin

What the plants are saying

Listening to plants could cut the amount of pesticides used on crops.

Plants aren't very good listeners but they might become good talkers with the help of two research teams at University College Cork. The €2-million study involves plant scientists and electronics specialists and has developed novel ways to get plants chatting.

"The whole idea is to use technology to monitor crops," explains UCC's professor of botany Alan Cassells. "We want to communicate directly with the plants." Your common or garden bean plant might not seem like much of a conversationalist but nothing could be further from the truth, provided you have the right ears for listening. Plant temperature can vary as a function of ambient temperature but also as a result of drought-related stress. Plant colour also gives off subtle clues about nutrient availability.

And 'green leaf volatiles' are a particularly rich source of information about a plant's relative condition, Cassells says. These are complex volatile chemicals that rise off the surface of a plant as a result of stresses such as insect or fungal attack. In some cases the volatiles have evolved as attractants that call in predator insects. Aphid damage releases volatiles that signal 'free lunch' to passing ladybirds.

The trick is being able to read the various clues being given off by the plant, Cassells says. "If you can detect these chemical signals then we can activate a response to the infestation. We are talking about sensors that can sense heat, greenness and volatiles." UCC's National Microelectronics Research Centre, now part of the Tyndall National Institute, leads the project with support from Cassells's Department of Zoology, Ecology and Plant Science.

The Institute is responsible for developing the sensors and hardware needed to respond to the plant signals identified by Cassells's team. Other partners include the Computer Technology Institute of the University of Patras, Greece, where the software to read the signals from the sensors is being developed, and the Eden Project in Cornwall, England. It will demonstrate the concept to the public and later to the trade.

The object isn't to make things easier for the farmer, Cassells explains. It is about reducing unnecessary and expensive farm inputs such as irrigation, fertilisers and pesticides. "Essentially, this is about sustainability because it is very difficult to sustain an agricultural system without inputs," he says. The traditional approach was to drench the entire farm with water if the soil felt dry, blitz all the crops with insecticides at the first sign of bugs and dose everything with nitrates if any of the plants look a bit yellow. The researchers want to reach a situation where signals from the plants themselves will indicate

when and what part of a field might need extra water. Only those sections of a field affected by, say, aphids will receive insect sprays and fertilisers are only applied when necessary.

Sensor placement is a key issue, he says. There is little point sticking moisture sensors in the ground if tractors can't then drive over them to reach crops. Equally, direct attachment of sensors to the plant isn't the best as they can become dislodged and only note what that one plant is saying. Rather, the research team is using 'proximal remote sensing' to find out what the plants are talking about. Cassells says a typical approach would be to have an array of specialised sensors attached to the boom arms of a tractor-borne sprayer. These would take the plant's analogue signal, convert it to digital and feed it back to the computer software, the 'black box' that will interpret the signals and decide what to do. "The black box is how the signals are interpreted to activate an appropriate response," he says. This response would be relayed back to the sprayer booms that would then apply water, pesticide or fertiliser as needed, but only where it was needed and not across the entire field.

We won't have to wait long for an opportunity to hear what plants have to say for themselves. The technology is already well developed and some demonstrations of the technology have already been carried out in Cork, Cassells says. More demonstrations are planned for later this spring at the Eden Project.

Thursday, 05 May 2005

✚ **UCC Department of Zoology, Ecology and Plant Science:** http://www.ucc.ie/academic/zeps/
UCC National Microelectronics Research Centre: http://nmrc.ucc.ie/
University of Patras Computer Technology Institute:
http://www.hoise.com/vmw/vmwc/articles/vpp/BOURA-PP-01-99.html
The Eden Project: http://www.edenproject.com/

Clocking all programs

Timing is everything in love and war, they say, but a researcher in Cork has taken timing to new levels. He devised a way to predict how long it takes for a computer program to run, something of crucial importance if the program is flying a passenger aircraft.

A Cork researcher has developed a program that can check the speed of other programs.

"The main issue is to improve software timing," says University College Cork computer science lecturer, Dr Michel Schellekens. "The industry wants to know when certain processes in the computer end. You need to be absolutely sure the activities end at a given time."

Schellekens is director of Ceol, UCC's Centre for Efficiency-Oriented Languages. He and his team have devised a novel computer algorithm that allows him to calculate how fast a computer program can get a job done. There is no room for approximations for some computer programs, he explains. He mentions aircraft autopilot software but other time-critical examples include the program that controls anti-lock braking systems on the family car. In these cases, the software has only fractions of a second to make a decision and initiate an action. Now Schellekens's algorithm can tell software designers how long the information processing will take. Before now programmers only had approximations for the longest time required to run a program to the end of an action. The new algorithm gives a more realistic average time no matter how complex the computer computation.

Software basically consists of a series of operations, where each operation takes a piece of information, transforms it and passes it on to the next operation for further manipulation, he says. In a sense, the information flows through the computer like water flows through a pipe. A computer can produce a list of the names of people on a given flight. If there are three people on the flight, the computer can order the names in six possible ways.The possible combinations grow rapidly as the number of passengers grows and by the time you reach 11 passengers, producing all possible name lists has already reached the processing limits for most personal computers, Schellekens says.

For larger data sets there are too many possibilities to get good averages of processing time required, even if you run test data, he says. "You can't run through all of them. You need a mathematical tool to predict it." Current computer timing systems are an average based on the 'factorial' number, the number of items in the set. The 11 passengers are 11 factorial, he says. "If you go to 20 factorial you won't be able to compute that any more. That is the problem people face in calculating average time of a software run," he says. If there were 65 passengers in a typical mid-sized aeroplane, that would produce more unique name lists than there are atoms in the universe.

Trying to produce a useable average of how long processing this information would take is a challenge with these large data sets. Rather than averaging the time taken for each permutation, Schellekens and his team developed a timing equation that can anticipate the average time needed to run a software program. "This algorithm can calculate in any circumstance. It is comparable to Boolean algebra." He calls his timing program Acett, average-case execution time tool, and late last month he had a successful test run to prove that it works.

Acett didn't need to watch the program running, it could derive running time directly from the program's own code in a "modular way." "Modularity basically means that there is great control over the predictability of the system, in that the behaviour of the entire system can be derived from the behaviour of its basic components. This is at the root of engineering. To predict a system, one analyses its parts and derives the behaviour of the whole. Achieving it for

software timing was a great research challenge which has been overcome at Ceol."

Acett would be of interest to real-time programming language developers, he believes. It could be used for any company interested in guaranteeing that its software meets deadlines, such as the aviation and automotive industries, chemical plants and robots.

+ Dr Michel Schellekens

"We are going to experiment with that and see if we can improve it," he says. He hopes to do this both by refining Acett but also by developing new types of microchips that are hard-wired to run Acett. "We are aiming to develop a whole new processor based on that idea."

Wednesday, 08 June 2005

+ **UCC Centre for Efficiency-Oriented Languages:** http://www.ceol.ucc.ie/

Bringing new life to old trees

Researchers at Teagasc and TCD are discovering innovative ways to rejuvenate our ancient trees.

Ireland has a small and diminishing collection of giants—not characters from a panto, but veteran trees. Efforts are now under way, however, to give our veterans a new lease of life and in the process protect this valuable genetic resource.

Teagasc's Dr Gerry Douglas of the Kinsealy Research Centre heads the project that also involves Trinity College Dublin researchers led by botanist Dr Trevor Hodkinson. "Our main object is to conserve the very few veteran trees we have in Ireland to give them another lease of life because we have so few of them," says Dr Douglas. "The second object is to find out why they are so special. They are unique because they have lived so long and they may have genes that confer longevity."

A veteran in this case is a tree that has a circumference greater than four metres when measured a metre or more above the ground. Many of our veterans have important cultural connotations, he says. The Brian Boru Oak in Tuamgraney, Co. Clare, is almost eight metres in girth and is named after our one and only high king of all Ireland, according to Dr Douglas. "The veteran sycamore at Kilmore Cathedral, Co. Cavan, is seven metres in circumference, is hollow, is very healthy and has an estimated planting date of 1631. It was recorded by and named after the local Bishop Bedell, who was the first scholar to translate the bible into Irish," says Dr Douglas.

These trees also have very tangible importance, he believes. "Our veterans are a rare genetic resource. They are a living link to past generations of trees, which migrated naturally here from the Costa del Sol after the last Ice Age. The oldest veteran oaks in Ireland are about 600 to 1,000 years old. It is sobering to think they may link back just eight or 10 generations to those original colonisers."

The researchers are interested in the unique characteristics that allowed these remarkable trees to survive. "All trees would not grow to such a great age," Dr Douglas says. Seed lines scattered by trees die out. Many trees that could become veterans are cut down in their prime for timber, so there is some luck associated with becoming a veteran. "For the ones that really live to be an old age, despite the environmental conditions, there is a genetic component to their longevity," Dr Douglas believes, and the Trinity College connection will hopefully reveal some of these trees' genetic secrets. Hodkinson and colleagues will do genetic fingerprinting of trees near the veterans to see if any genes have been passed across from the ancient trees given that many of them are still sexually active. "We are not specifically looking for genes that give longevity. We want to see if the veterans have contributed to any of the trees in the surrounding area." Dr Douglas notes, however, that 'Methuselah' genes conferring long life have been identified in other species, including microscopic worms and fruit flies. It is not inconceivable that trees might also have genes that predispose to long life.

The Kinsealy research group is involved in the conservation of the veterans. "We can extend their lives for another generation by propagating them vegetatively," says Dr Douglas. They are concentrating on oak, ash, sycamore and elm and have collected shoots from the old trees. These in turn have been grafted onto fresh seedling root stock that has been species matched. The elm is an exception, and will be grafted onto elm root cuttings from the veteran, producing a true clone. The team hopes that survivors of the devastating Dutch elm disease may have inherent resistance to the scourge. There is no option with the other three species but to make use of grafting, as their roots will not take. "It is the only way we can propagate them," says Dr Douglas.

This year's grafting efforts have delivered promising results, with viable plants produced from oak, ash and sycamore. Significantly, the shoots have grown by a

metre, much more than they would have had they remained on the veteran tree, says Dr Douglas. It may be that the old shoots are learning to behave like young ones after association with young roots. "The grafted plants will be returned to the original tree owners at the end of the summer," he says. "They will then plant them out and so the life of each veteran tree can be guaranteed to continue in a rejuvenated form." Additional cuttings will also go to the John F. Kennedy Arboretum in Wexford in order to create a collection of veteran trees.

Very old trees, including the veterans, provide a largely hidden service on behalf of biodiversity. Their roots, bark and often rotting interiors support a remarkable number of other species, some of which now face extinction due to the diminishing number of old trees in our landscape, Dr Douglas says. They are known as home to fungi, lichens, mosses, a host of invertebrates, microbes and slime moulds, he says. "The more obvious inhabitants: bats, squirrels and birds may have taken up residence too." Once the bark of a very old tree has been breeched it succumbs to wet rot, but this too is of benefit to many other species. Some are highly specialised and depend on veteran trees and old trees to reproduce.

"Sadly we know that many of the specialised beetles and fungi found in veteran trees are in danger of extinction," says Dr Douglas. "There are 447 fungi listed as endangered in the British Red Book, and 400 of these are most readily found in veteran trees and woodland. Similarly for beetles, we know that at least 12 old woodland species have become extinct since the Bronze Age."

Thursday, 14 July 2005

➕ **Teagasc Kinsealy Research Centre:** http://www.teagasc.ie/kinsealy/contacts.htm
TCD Botany Department: http://www.tcd.ie/Botany/

+ Brian Boru Oak, Co. Clare

Chinese hamster in great demand

Hamster stem cells aid production of protein useful for medical treatments.

The Chinese hamster has suddenly become very important to Dublin City University. A research group there has joined with pharmaceutical giant, Wyeth, in a €4-million research collaboration focused on cells cultured from this small, furry mammal.

The project received not one but two launches last month, one at DCU and another at the big Bio 2005 Convention, held this year in Philadelphia in the US. The deal is highly significant given that it involves very advanced research in Irish labs on behalf of a huge international biotech player.

"This is actually embedding industrial research in Ireland," says the director of DCU's National Institute for Cellular Biotechnology, Martin Clynes. "The American research arm is taking Ireland seriously. It really has been a success story by SFI and the HEA," he adds, referring to Science Foundation Ireland, which put up the funding, and the Higher Education Authority's investment in facilities at DCU under the Programme for Research in Third-Level Institutions. Wyeth has invested more than a billion euro here to build the largest single biotech-based pharmaceutical production plant in the world. DCU now has links with research collaborators at the Grange Castle, Clondalkin, plant and other Wyeth scientists based at Andover, Massachusetts.

The project focuses on an important individual in the biotech field, the Chinese hamster. Cells taken from the hamster's ovary have become one of the most widely used cell types in biotech plants. "They are the primary cell used for biotechnology production," says the director of cell and molecular services at Andover, Tim Charlebois. The cells are first cultured and then engineered by inserting an extra gene, one that produces a valuable pharmaceutical protein useful for medical treatments. The cells are cultured in large tanks and then the target protein is extracted from the cells and purified.

"It is a four-year collaboration for the purpose of doing research to improve the production of mammalian cells for biotechnology," says Charlebois. "The hamster cells are more efficient than yeast or *E. coli*. The proteins they are making are very complex and mammalian cells are better adapted for that purpose." The problem is, Wyeth has hit production limits using this method, making these important protein-based

pharmaceuticals expensive but also forcing them into short supply.

The goal of the project is to find ways either to increase the amount of protein the cell can produce or increase the numbers of cells that can be cultured in a production tank. "We think we are going really well but I believe we have a long way to go to improve the system," says Charlebois. The research will study the biochemistry of the hamster cell in minute detail, looking in particular at its gene expression as it grows in the tanks. "We are hoping to study the biology of the cells under production conditions," he says. "We are trying to understand its fundamental biology."

The work could bring about profound changes to the cell itself, suggests the development director at Grange Castle, Brendan Hughes. "We want to look at what happens and what goes on under different conditions and then go in and modify the cell's genome," he says. It all comes down to how the cell responds to the production conditions it finds within the tanks. "We will have eight people working on it here," says Clynes. "We will be looking at gene expression, at the RNA and protein level and at a variety of cell types given to us by Wyeth." Some engineered cells work better than others for unknown reasons, he says. "We are going to try to work out the molecular reasons for this. We may be able to engineer the cells to make them perform better."

Thursday, 21 July 2005

✚ **DCU National Institute for Cellular Biotechnology:** http://www.nicb.dcu.ie/
Wyeth-Andover, Mass.: http://www.wyeth.com/divisions/Andover.asp
Wyeth-BioPharma Campus, Clondalkin, Dublin: http://www.wyeth.ie/biopharma.asp

Looking for a fallen star

A Dublin institute is to the forefront in the study of the 'failed star' known as the Brown Dwarf.

Researchers at the Dublin Institute for Advanced Studies have made important new discoveries about the small, indistinct 'failed stars' known to astronomers as a Brown Dwarf. Their findings have revealed a great deal about these objects, how they form and whether they are more like a small star or a big planet.

"People really don't understand how these Brown Dwarfs form," says PhD student in the institute's School of Cosmic Physics Emma Whelan. They are small, visually weak and very difficult to study as a result, she says. That is what makes the findings by Whelan and her supervisor at the Institute, Prof. Tom Ray, so important. Their discovery, using telescope data recorded by collaborators at the Osservatorio Astrofisico di Arcetri in Italy, was published last month in the journal, *Nature*.

"Most of what we know about how stars form is from low-mass stars," says Whelan. Star birth occurs inside huge clouds of dust and debris such as the Giant Molecular Clouds of our Milky Way galaxy. The dust begins to collapse in on itself due to gravitational pull. If enough mass accumulates there may be sufficient heat and gravity to trigger the hydrogen fusion that typifies all true stars. More dust circulating around the star may also go on to collapse into circling planets.

Brown Dwarfs form when too little material has accumulated to trigger hydrogen fusion, Whelan says. These failed stars are intermediate between a low-mass star and a oversized planet, with a typical Brown Dwarf having a mass of between 20 and 70 times that of Jupiter. "The most important question relating to these objects must be how exactly do they form, more like stars or more like planets," says Whelan. "This puzzle has repercussions for the theory of star formation as a whole."

One clue to this puzzle is seen repeatedly in 'baby' stars, protostars that have only just formed. They all start by accumulating mass from the surrounding molecular clouds, but counter-intuitively they all also eject powerful outflows of fast-moving material back out towards the cloud. The disc of matter that will go on to form a protostar is rotating and some of the infalling matter is launched back out along the magnetic field lines surrounding the protostar as a result of centrifugal force. These jets or outflows produce huge, fast-moving streamers, measured in parsecs ($3 \times 10 < MD +> 1 < MD +> 3km$) and speeds measured at tens of kilometres a second.

These jets have an important purpose. They take angular momentum away from the rotating protostar, slowing it and thereby allowing it to continue pulling more matter into the forming star. The jets are also thought to blow back the surrounding dust cloud, allowing the star to emerge into the open. This outflow mechanism was known to exist in low-mass and very large stars and also in those gravitational monsters, Black Holes. "People wondered if Brown Dwarfs had outflows as well but they are small and are visually weak," says Whelan.

Whelan and Ray answered this fundamental question with their study, however. They used recorded spectrographic data captured by a large telescope in Chile by the Italian team. The Dublin researchers used a technique known as 'spectro-astrometry', a method that allows the measurement of distances from spectroscopic data. "It is the only way to see the outflow from a Brown Dwarf," says Whelan.

The data was for Brown Dwarf rho-Oph 102, located in the rho-Ophiuchi molecular cloud. They managed to identify an outflow from the body moving at about 40km per second and reaching out a full 15 times the distance between the Earth and the sun, or

+ Emma Whelan and Professor Tom Ray

about 2,244 million kilometres. "It is quite a big deal," says Whelan. The first proof that Brown Dwarfs can produce outflows may provide answers to a number of questions. The finding adds weight to theorists who hold that Brown Dwarfs form in much the same way as low-mass stars, she says.

More significantly, it also shows that something as small as a Brown Dwarf can produce outflows using the same mechanism as the giant stars and Black Holes, despite the 10 orders of magnitude size difference between these bodies. "It is the same mechanism irrespective of mass," says Whelan. "This important mechanism is thus very robust and it does not now seem impossible that outflows may also accompany the formation of planets. The idea that perhaps Jupiter or Saturn once were the drivers of such impressive flows is indeed tantalising."

Thursday, 21 July 2005

⊞ **DIAS, School of Cosmic Pyhsics:**
http://www.dias.ie/index.php?section=cosmic&subsection=index
Osservatorio Astrofisico di Arcetri: http://www.arcetri.astro.it/
Emma T. **Whelan**, Thomas P. **Ray**, Francesca Bacciotti, Antonella Natta, Leonardo Testi and Sofia Randich, (June 2005) A resolved outflow of matter from a brown dwarf, *Nature* 435, 652–54
Nature: http://www.nature.com/nature/journal/v435/n7042/abs/nature03598.html

+ Dr Emma Teeling of UCD with an example of a myotis species bat in the Biology Department at UCD. Prior to publication of her research in *Nature* it was assumed that of the two main families of bats—the non-echolocating, fruit-eating megabats and the smaller, omnivore, echolocating microbats—the megabats must have evolved first. Researchers believed that megabats would not have discarded this valuable mutation and that echolocation must have been a later evolutionary development. Teeling and her colleagues' analysis changed all of this, however. The study revealed that in fact the microbats evolved first.

+ Dancers take to a crowded, smokey dance floor

Effects of ecstasy now considered to be even worse than feared

Ecstasy damages the brain but also has toxic effects on muscles, according to new research at UCC. The muscle tissue can't take the sustained stimulation and begins to break down.

Ecstasy is a constant risk to recreational users given its unpredictable impact on the brain and nervous system. Now researchers in Cork have discovered that the drug can also cause muscle tissue to dissolve.

"Ecstasy should be regarded as a toxic chemical and not as a recreational drug," says professor of medicine and biochemistry at University College Cork, Prof. James Heffron. He has good reason for stating this given the effect it has on skeletal muscles. He and colleague Prof. Frank Lehmann-Horn at the University of Ulm have discovered that ecstasy can act on tissues outside the central nervous system. "It was said to have central effects only but we have shown it does also have effects in the periphery," says Heffron.

The drug, also known as MDMA, is very popular among young people attending clubs, but the synthetic amphetamine continues to cause unexpected and tragic deaths. It produces psychedelic and mood-changing effects by acting on the central nervous system, but MDMA also triggers toxic reactions. One of the most common, a rapid and difficult to control rise in body temperature, is the reason Heffron became involved in the research, which has been released on the internet by the *Journal of Pharmacology and Experimental Therapeutics* and will be published in the *Journal* next month.

"It came about because we were working on diseases that lead to high temperatures, particularly malignant hyperthermia," says Heffron. Malignant hyperthermia (MH) arises in rare cases when surgery patients react badly to anaesthesia. Their body temperature rises rapidly, and can cause death if not controlled. "We thought that people who would be susceptible to MH, who have the gene for it, might be susceptible to death from ecstasy," he says. "It turned out that that was not the case," he says. Yet one very valuable out-turn from the research was that the team uncovered the drug's unexpected impact on voluntary muscle tissues. "We showed how ecstasy works in the body," says Heffron.

All of the symptoms linked to MDMA use, including muscle contraction, pain and spasms and body-temperature rise were thought to be linked to the drug's interaction with nervous system tissues. Heffron and Lehmann-Horn's work showed that the drug can also work outside the central nervous system, interacting with specialised receptor proteins on the surface of muscle cells known as nAChRs. These are found at the junction between nerve and muscle tissues, the myoneural junction. The ecstasy or some component of it attaches to the receptors, switching on the muscle and over stimulating it. "It triggers the muscle to go into contractions and leads to heat production," explains Heffron. "It leads to a breakdown of the muscle. It is literally breaking up the tissue, the tissue dissolves."

The muscle tissue can't take the sustained stimulation and begins to break down in a process known as rhabdomyolysis. This release of cell contents also leads to the severe toxic effects of the drug outside of the central nervous system. The kidneys for example are affected by this. "There are well known renal effects," says Heffron. "That would probably be a consequence of the muscle breakdown. Material from the muscle tissue clogs up the filtering system of the kidney." The situation for users may be complicated by the fact that street ecstasy pills usually contain a variety of other compounds, including caffeine, that may exaggerate the actions of ecstasy itself, Heffron adds.

Potential benefits from the research may be opportunities for an antidote to help those who react badly to the drug. It may be possible to develop a substance that can either block ecstasy from latching onto the nAChR receptor sites on the muscle or unseat them if they are already in place.

Thursday, 18 August 2005

⊞ **UCC School of Medicine:** http://www.ucc.ie/acad/medschool/

UCC School of Biochemistry: http://www.ucc.ie/academic/biochemistry/

Werner Klingler, James J.A. **Heffron,** Karin Jurkat-Rott, Grainne O'Sullivan, Andreas Alt, Friedrich Schlesinger, Johannes Bufler and Frank Lehmann-Horn, (September 2005) 3,4-Methylenedioxymethamphetamine (Ecstasy) activates skeletal muscle nicotinic acetylcholine receptors, *Journal of Pharmacology and Experimental Therapeutics* 314, 1267–73.

Journal of Pharmacology and Experimental Therapeutics:
http://jpet.aspetjournals.org/cgi/content/abstract/314/3/1267

The expressive dress that shows how women really feel

These are no ordinary ball gowns. One includes sensors that detect the heart as it pulses and matches the beat by lighting up a bright red panel across the front of the dress. The faster the beat, the faster the panel flashes, lighting up like a neon sign.

The latest in high fashion party wear has a little something extra to offer this year. A Dublin-based research group has developed a stylish dress that looks good but also tells you how the woman wearing it feels about you.

A second gown samples the wearer's 'alert' response seen in changes to her 'galvanic skin response', how well the skin conducts electricity. The more impact a male suitor has, the higher her alert status, and this raises—or lowers if you are not doing well—the brightness of spheres suspended in the garment's outer layer. These gowns have the potential to take some of the mystery out of human relations but the goal is not to win hearts but to develop wearable electronics.

A number of dresses that combine fashion with electronic signal processing were developed by the Adaptive Information Cluster (AIC). The Cluster is a multi-disciplinary research group involving senior researchers from Dublin City University's Centre for Sensor Research and School of Computer Applications, and University College Dublin's School of Computer Science and Informatics, explains the head of the UCD school and an AIC principal investigator (PI), Prof. Barry Smyth.

The Cluster received Science Foundation Ireland funding worth €7 million and brings together the work of five PIs, says Smyth, "each of them a professor in either UCD or DCU." It combines the efforts of scientists working in disparate areas, including sensor science, software engineering, electronic engineering and computer science. Headed by DCU's Prof. Dermot Diamond, the AIC started up about two years ago and involves the work of about 70 researchers, says Smyth.

PhD candidate at UCD Lucy Dunne carried out the research that delivered a number of 'expressive garments' that include sensors that read the wearer's mood. They detect

+ No ordinary ball gowns. Each dress responds to its wearer's reactions to suitors.

The trick is knowing where to put the sensor to get useable information, she adds. While the four gowns she designed, two of which have been installed as part of a display at the Brigham Young University Museum of Art in Utah, US, focus on individual expression, "most of my research is about function."

Wearable electronics could deliver many useful functions, particularly in the medical area. People with limited movement could activate switches or direct motorised wheelchairs using this technology. Dunne has also produced a prototype sports jersey that monitors heart rate and respiratory effort and the technology could be built into trainers to monitor pace and gait.

Thursday, 1 December 2005

pulse rate, galvanic skin response, the 'startle' response and whether the person is laughing, explains Dunne.

The gowns are smart in two ways, looking well but also capturing real-time information on a person's physiological condition. They incorporate sensor technology developed at DCU with electronics, but must still be able to go into the washing machine after the ball and come out working.

These are no ordinary sensors, she says. "It is a really nice, crushable foam and integrates well into clothing and is washable." Changes in the foam's shape produce changes in its electrical response, generating a signal that the smart dress interprets.

+ **The Adaptive Information Cluster:** http://www.adaptiveinformation.ie/home.asp
National Centre for Sensor Research: http://www.ncsr.ie/
DCU School of Computing: http://www.dcu.ie/computing/index.shtml
UCD School of Computer Science and Informatics: http://csiweb.ucd.ie/

A spoonful of sugar

Your house is probably full of sweets and Christmas pudding and lots of sugary things, but they won't include the kind of sugars being made by Dr Paul Murphy. The University College Dublin chemist is synthesising the specialised sugars that support vital communication links between living cells.

Research into complex sugars involved in cell communications could have therapeutic value.

Murphy is a senior lecturer in the School of Chemistry and Chemical Biology at UCD and is a principal investigator at the ¤26-million Centre for Synthesis and Chemical Biology. Funding came from the Higher Education Authority's Programme for Research in Third-Level Institutions.

The new centre involves the work of 37 research investigators from UCD, the Royal College of Surgeons in Ireland and Trinity College Dublin. "In the centre we work on wholly new molecules, new compounds that are interesting in terms of their biological properties," Murphy explains.

The challenge is to isolate a molecule of biological interest and then find ways to duplicate or synthesise the molecule so it can be tested for its biological effects. "People are isolating a lot of compounds from natural products and doing analysis of their biological action. Of small molecule therapeutics, 50 to 60 per cent come from natural products or substances closely related to natural products," says Murphy.

His own work focuses on sugar molecules. They are true sugars, carbohydrates, but are very complex compared to what most of us understand as sugars. "Everyone has heard of glucose, but these sugars are more complex," he says. "They coat cells and are involved in communications between cells. These sugars also mediate recognition between bacteria and their cell hosts."

Researchers are interested because of the therapeutic potential associated with them, Murphy says. Cuban researchers last year developed one of the first carbohydrate vaccines against meningitis and there is great potential for many more. Scientists have been slow to exploit this potential, however, because the technology

+ Dr Paul Murphy, senior lecturer in chemistry and chemical biology at UCD, photographed with researchers in carbohydrates, from left, Wayne Pilgrim, Guillaume Anguetin, Jerome Lalot, Christina Loukou and Dearbhla Doyle.

wasn't there to do so. There are machines that can now copy out sections of our genetic blueprint, DNA, and produce small biomolecules called peptides, but synthesising complex sugar molecules is very difficult, he says. "The technology hasn't been there to deal with the carbohydrates."

DNA and peptides, short chains of amino acids, are very regular structures, he says, but these sugars can have "huge diversity" even though they produce the same biological effect. "To produce a pure glycoprotein is very difficult." The body has no difficulty synthesising the complex sugars it needs and relies on enzymes, the body's own nanomachines, to install sugar components on proteins.

Murphy received a €1.2-million award from Science Foundation Ireland to fund research looking at angiogenesis, the process that forms blood vessels. Sugars seem to moderate angiogenesis and have been shown to inhibit the development of endothelial cells, the cell type that lines the insides of blood vessels.

He is also looking at "imino sugars" which are similar to glucose but have a nitrogen in

the molecule replacing an oxygen. They have a range of interesting biological effects, including being able in some cases to inhibit enzyme activity. They can also imitate peptides despite being completely different in form and shape, Murphy explains. An imino sugar structure is created and then small amino acid 'side chains' are attached. When presented to a cell the imino sugar performs just like the original peptide but is much more stable and longer lasting because of its tough sugar-based structure.

It has been a good year for Murphy who earlier this year won an Astellas USA Foundation Award. Recipients are chosen after nominations from internationally-distinguished scientific colleagues, and then are invited to make a formal application to the foundation.

A small financial award comes with the prize, enough to support the work of a PhD student who will work with Murphy on carbohydrate chemistry in the Centre.

Thursday, 22 December 2005

+ **UCD School of Chemistry and Chemical Biology:** http://chemistry.ucd.ie/
Centre for Synthesis and Chemical Biology: http://chemistry.ucd.ie/cscb.html

+ Professor Jochen Prehn

Studying cell suicide

Which genes dictate the life and death of cells? A team led by Professor Jochen Prehn hopes to find out.

Individual cells know when it is time to go and they initiate their own spontaneous death. This process can go dangerously wrong in a number of diseases, however, and a Dublin research group hopes to discover ways to control this.

The Royal College of Surgeons in Ireland (RCSI) runs a very active research group studying the cell suicide process, known as apoptosis. It involves a team of 20 to 25, explains the head of physiology and medical physics at RCSI, Prof. Jochen Prehn.

"The goal is to identify the link between cell stress signalling and cell shutdown," he says. "We find there is a fine balance between cell survival or death when under stress."

Death is a part of life when it comes to the cells in our bodies. Unwanted or ageing cells undergo apoptosis as a highly controlled and safe way to dispose of themselves. Very complex signalling is involved to initiate this process, and Prehn's group studies both the signals and the expression of proteins that begin the apoptotic cascade of events leading to cell death.

Things go wrong in diseases such as cancer, however, where cells refuse to die on cue. Others include Parkinson's disease and motor neuron disease, where cells go into apoptosis too soon and central nervous system cells are lost too quickly.

Prehn came to Ireland in July 2003 to take up a Science Foundation Ireland research professorship at RCSI. He was already a noted researcher specialising in death signalling in neurons and he has continued this work here. Teams within his group are working on epilepsy, on damage caused by stroke and on motor neuron disease. "We are trying to identify the genes that dictate cell survival or death," he explains.

The research is conducted both *in vitro* and *in vivo*. Cell cultures derived from mouse hippocampus cells are used to study the up and down regulation of proteins inside the cell after stressing the cells. In this way they can watch what the deciding factors are between cell survival and death. They also use a mouse model for motor neuron disease that closely matches the progress of the disease in humans, but with a disease process lasting 15 to 20 weeks rather than years. "We can see how the genes behave *in vivo*. There is a great need for *in vivo* models for these diseases," Prehn says.

Microarrays and related technologies are used to sample the proteins being expressed after a cell has been stressed, for example through reduced blood flow. "If we know a certain protein is important, for example in ischemia, we know there is misfolding of protein structure under stress," says Prehn. They can watch for proteins that help to tip the balance between cell survival or death, and if found, this discovery opens up opportunities for new drug therapies, he explains.

If they can identify the protein pathways for proteins that improve survival, then perhaps drugs can be found that enhance this effect, he says. "We try to boost the cell survival response and by doing this you can make cells more resistant even after the onset of disease. We want to identify ways that signalling can be boosted."

It also works the other way, particularly in cancer. "In this case we want to find treatments that sensitise cells to apoptosis." Drugs in this case would encourage cancer cells to relearn the rules and become susceptible to apoptosis again.

Thursday, 5 January 2006

➕ **Royal College of Surgeons in Ireland**: http://www.rcsi.ie/

Finding the decisive factor

Those having difficulty choosing this year's summer holiday might consider talking to a research group at University College Cork, where computers with artificial intelligence are making important decisions.

Can't make your mind up? A research group led by Professor Gene Freuder is developing technology that delves into how we make important decisions.

The Cork Constraint Computation Centre (4C) develops computer software and the underlying science to help businesses and individuals make good decisions, explains 4C director, Prof. Gene Freuder. "Our motto is 'making hard decisions easier' and that is what we try to do," he says. The idea is to solve complex decision-making problems using computers 'taught' to weigh up the pros and cons while including the constraints operating on any given decision.

Freuder came to Ireland in 2001 from the University of New Hampshire and is a professor in UCC's Computer Science Department. He holds a €7.5-million research-fellow grant from Science Foundation Ireland and leads a research team at 4C that involves up to 50 people. The centre has also attracted support from Enterprise Ireland, from the Irish Research Council for Science, Engineering and Technology and from industrial partners. Freuder made news in late October when he was elected as a Fellow of the American Association for the Advancement of Science (AAAS). He was one of only 14 new Fellows elected in 2005 for the information, computing and communications section, out of a total of 376 new AAAS Fellows.

"I was very pleased, and pleased for the Centre and hope it reflects well on Science Foundation Ireland and its choices," he says of the distinction. The award certainly arose as a result of the work he has done for some years in artificial intelligence (AI) and constraint computation. "I come from an AI background so I use inference and other AI techniques. One of the active research areas is integrating AI and OR [operations research] techniques to solve problems."

We make hard decisions every day but don't really think about what goes into making a decision. "There are all sorts of choices we have to make with lots of decision variables and constraints," says Freuder. "It is difficult to make decisions under these circumstances. We work on the underlying science and the applications needed to help people make better choices. Constraint computation allows you to do that sort of thing but on a much more sophisticated level."

As an example, he described the decision-making involved with a person shopping for a digital camera. "At the most basic level you might want to know what is possible," he says. Can you buy a digital camera for less than €20? Or can you get a moderately good camera for under €100? Are you satisfied with a one-megapixel camera or do you need five-megapixel resolution? If so, how much are you willing to pay if it is more than €100? The computer weighs up the options while looking at what Freuder describes as the 'hard' and 'soft' constraints. His researchers develop mathematical algorithms that allow a problem to be solved in a computational way.

4C conducts extensive research but has also established links with companies here to work on real-world problems. Projects have included work on supply logistics with Cork University Hospital and on the optimisation of floor plans at Bausch & Lomb in Waterford.

"We want to make it easier to use so it will have more of an impact," says Freuder. Some software is specifically designed for a specific company problem, but many of the algorithms can be applied in other situations so there is the potential for 'off the shelf' computer software in support of easier decision-making. One project involved developing a system to look at holiday travel packages. There is no end of applications for this type of computerised decision assistance for those who don't mind letting a machine help out.

Freuder's research team has grown quickly and includes academics, post-doctoral researchers and graduate students. It also has a strong international dimension. "We have brought in people from all around the world and I think we are building something nice here."

Thursday, 5 January 2006

➕ **4C:** http://4c.ucc.ie/web/index.jsp
UCC Department of Computer Science: http://www.cs.ucc.ie/

$m'-1$

m

G' Lemma

b b c
c d d

a a
b
c
d

save big things reuse
bt free
Search

K-trees

Rina Dechter
width
AIS

Complies
Subr

Several
in
parallel SUSSMAN

not

+ Professor Gene Freuder

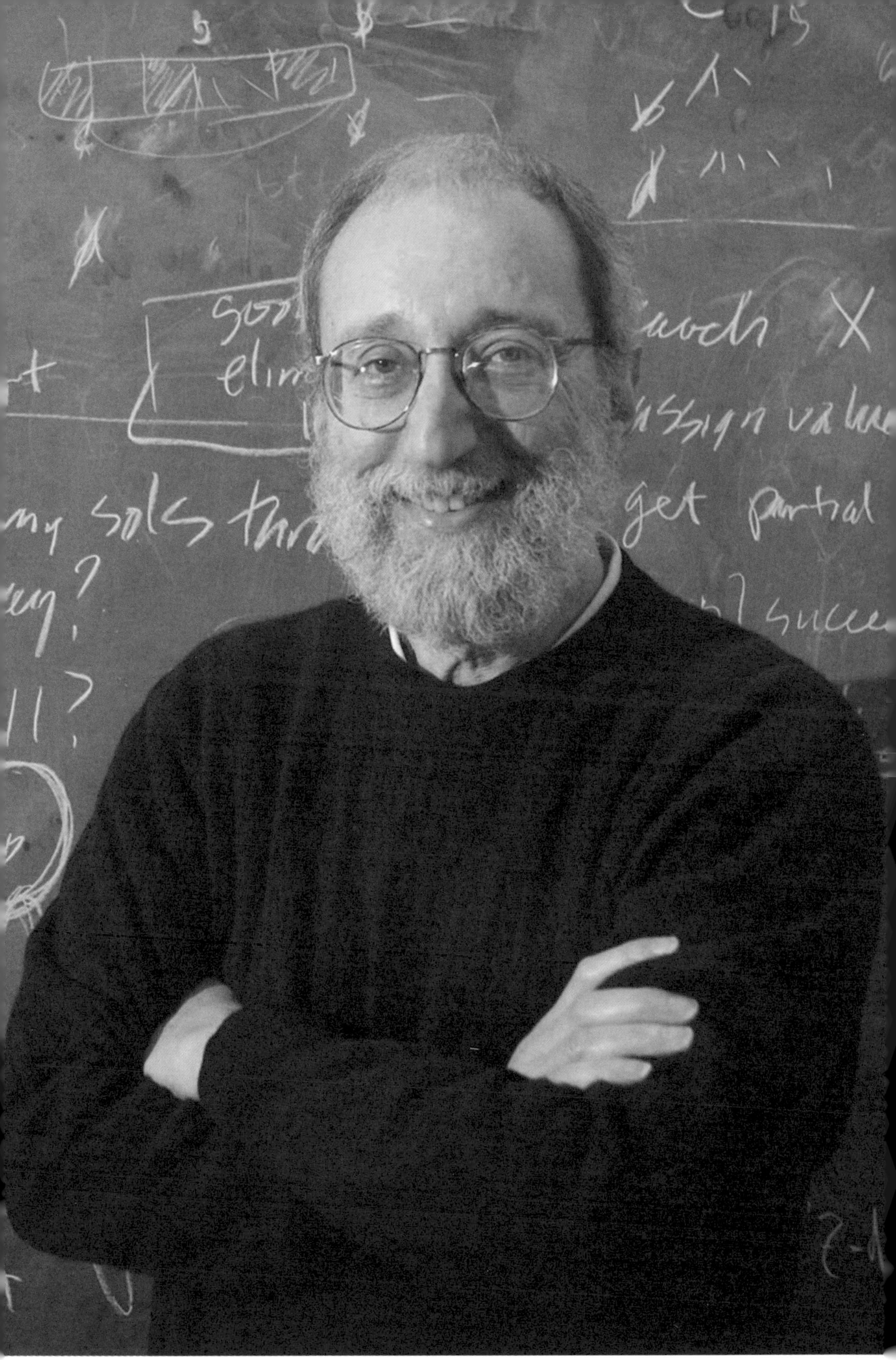
get partial

Linking life
to the web

Getting wired to the world takes on new meaning in a research initiative at Dublin City University where a group headed by Prof. Dermot Diamond wants to connect us physically into the world wide web.

There are immense technological hurdles but also many real-world benefits that could come from the research says Diamond, who is professor of analytical chemistry and director of the Adaptive Information Cluster, an initiative that partners DCU and University College Dublin.

"The theme I have been following for the last five or six years is pervasive communications and computing," explains Diamond who is funded by Science Foundation Ireland. The concept is based on the idea that computing power and the ability to communicate "will be available everywhere on demand."

He isn't just talking about voice, e-mail and images over the web. He pictures a situation where we will be connected in physically as well. "The data on the internet is dominated by what people are typing in or putting in as pictures or video. It is media dominated and relies on human input," he explains. "The next phase of this is where the pervasive web becomes wired to sensors because the range of information you can get into the web will be expanded."

You can view this as big brother watching your every move, but introducing 'body sensor networks' could be a boon to those who really need them, for example the elderly or those whose movement is impaired. Also known as 'ambient assisted living', it comes in for major research support under the EU's seventh framework programme. "The emphasis is to keep people out of hospital and improve their quality of life. It is being able to monitor people's activity and well being, their wellness," says Diamond.

His group is studying modified foams and conducting polymers that send signals when compressed or stretched. These could indicate whether an isolated individual was moving about through normal daily routines. Any sudden change to an individual's established routines would be flagged by the system, based on information coming in from wearable sensors on the individual.

+ Professor Dermot Diamond

"We are working with sports science people to develop a vest that monitors breathing," he says. You wear these sensors and they relay information over the internet using built-in communications. International groups have developed what are known as 'motes', tiny coin-sized remote sensors that have built-in communications, enabling them to communicate with each other in a network, but also to a central location.

Diamond's group has joined with University College Cork's Tyndall Institute to develop Irish made motes that can provide this networked capability. The idea is to go from dozens of sensors to millions of sensors, all networked and providing information about the environment, about the safety of food products and other 'real world' uses, says Diamond. The idea is to force down the cost of the motes through technological advances, "how do you make sensors cheap enough and small enough and reliable enough" being the question, he says.

His group is working on a food quality monitoring system that can track the temperature at which fish and shellfish have been stored from the time they are taken from the water until they are cooked. Another idea being studied is a dye in packaging that changes colour if a product begins to spoil. A reader detects the colour change and feeds this information to a central database.

The greatest challenge, however, is to get biological information directly into the computer, Diamond suggests. "How do we bring chemical information and biological information into the digital world? How do we bridge the molecular and biological world and the technological world?" This is currently done via laboratory equipment, but this equipment is large and isolated from the individual. "How do you get these measurements out of the lab and into the internet world," he asks.

Existing biosensors tend to be simple transducers and not true sensors, he says, thermisters for temperature, photodetectors for light. "They are not really sampling the molecular world. They tell you something about it but are not really connected to it." He pictures a situation where true chemical sensors will be able to connect directly to the individual to provide information about health status.

Thursday, 19 January 2006

➕ **The Adaptive Information Cluster:** http://www.adaptiveinformation.ie/home.asp

Magnetism in a spin

A TCD group is looking at the use of magnetic materials in an attempt to reduce computers to the size of a €2 coin.

Funny old thing magnetism, just when you think you understand what it is and what it can do, something new comes along. A Trinity College Dublin physicist has used it to levitate non-magnetic materials and to stop two liquids in a single container from mixing.

Professor of experimental physics in Trinity's School of Physics, Michael Coey is used to being surprised by magnetism and magnetic materials, having studied them for much of his professional research career. Advanced magnetic materials have allowed us to pack 100 billion bits of computer information on a single square inch of computer storage space. They also look likely to produce the next generation of computers, systems that won't have to be reloaded with software every time you switch them on.

Coey is a leading international expert in magnetism and new magnetic materials. He is ranked 309th in a list of the 1,100 most-cited physicists in the world, the only person working in Ireland to appear on this list. He is a foreign associate of the National Academy of Sciences in the US, and a Fellow of the Royal Society. He won the Royal Irish Academy's inaugural Gold Medal in the Physical and Mathematical Sciences last year, but Coey is also a leading light in the development of a multi-purpose public science gallery in the new building that will house Trinity's new CRANN nanoscience research laboratories.

In recent years, Coey and his Science Foundation Ireland-funded research group have focused their efforts on "nanoscale magnetics and spin electronics", he explains. "The broad question is what goes funny or what is different if we reduce our magnets to the 10 nanometre range." A magnet that size is almost impossible to imagine. A human hair is 8,000 times thicker, says Coey.

And yet all modern computer magnetic memory is based on nanoscale magnetics, he says. It relies on magnetism and the building up of layers of thin films, some conductors, some insulators and some magnetic. "It is quite astonishing that people are able to make these films one nanometre (one billionth of a metre) thick on a piece of wafer as big as an old record."

Spin electronics involves using the smallest natural magnets available—electrons. "The idea of spin electronics is that up to now we only made use of the electron's negative charge, while ignoring the fact that the electron is in itself a little magnet," Coey explains. "The big idea with spin electronics is to use the idea that the electron is a little magnet and see what we can do with it."

The approach is based on a phenomenon first seen in a French lab in 1988 but which, within a decade, had brought about the first generation of nanomagnets and spin electronics.

Coey's group studies new types of nanofilms using unusual combinations to create novel magnetic effects. "The concept is really very simple. The heart of the device is a sandwich with some magnetic layers and some space," he explains. "But it has to be built on such a tiny scale." The team's particular strength is its long experience with novel magnetic materials, something that informs what new combinations to try. He believes that this technology will become at least as important to computing as semiconductors. The real breakthrough will come when second-generation nanomagnetic materials actually replace existing semiconductors to allow the production of magnetic transistors. "What would be wonderful is if we had a magnetic semiconductor that could make spin transistors," he says.

His group has already developed unusual new magnetic materials using combinations that should not really be magnetic at all. "We are very interested in a new group of materials which, to everybody's surprise, were ordinary oxides that when doped with ferromagnetic material become magnetic at high temperatures. We don't understand why the magnetic interactions are so strong."

Meanwhile, the group continues to study novel things about magnetic fields, for example that you can make non-magnetic materials levitate in a magnetic field when immersed in paramagnetic liquids. Coey has also discovered that fields can prevent two liquids in a single container from mixing through convection. He can also use magnetic fields to induce mixing in two liquids, a technique valuable in a research area known as micro fluidics.

He is also looking at the use of magnetic materials linked to biomolecules, which could be used in completely new medical imaging technologies.

Thursday, 26 January 2006

⊞ **TCD School of Physics:** http://www.tcd.ie/Physics

CRANN (Centre for Research on Adaptive Nanostructures and Nanodevices): http://www.tcd.ie/Physics/Crann/

+ Professor Michael Coey of TCD physics department holds a piece of magnetised carbon from a meteorite that landed in Canyon Diablo in the US.

Future prospects

A second-year student at Kinsale Community School, Aisling used harmless bacteria in an inexpensive device that lets the consumer know when a food product has not been stored safely. The simple colour-change indicator is easy to read and is included in the food item's packaging, making it as simple as possible for the consumer.

Ireland's scientific traditions are in safe hands, as seen in the winner of the 2006 BT Young Scientist and Technology Exhibition. Top young scientist Aisling Judge, at 14 the youngest ever Young Scientist winner, impressed the judges with a biological system for detecting food spoilage.

"I developed a spoil indicator that indicates when food has gone off," Aisling explains. She wanted it to be as easy as possible for the person to use and the colour of the indicator shows whether food is fresh, deteriorating or spoiled. "It would be part of the packaging," she says. "It wouldn't cost enough to change the price of the food."

The judges liked the innovative approach taken by Aisling in her project, which was overseen by her teacher, Catriona Barrett. It involved the novel use of existing materials in a wholly new way. The system is based on the use of a harmless bacterium found in milk, *Lactococcus lactis*. Aisling developed a small test unit that contains the bacterium and also nutrients. It is sealed up so the bacteria can't escape and doesn't actually touch the food itself.

Included in the unit is an acid-sensitive indicator that changes colour as acid levels rise. If temperatures rise too high and bacterial growth is encouraged then the *L lactis* will begin to grow, using up the sugar and nutrients in the indicator. Their metabolic activity also causes acid levels inside the unit to rise, causing the unit to change colour.

Aisling ran a battery of tests from last November (2005), matching the growth of an important and well-studied harmful bacteria, *E coli,* against the growth of the *L lactis* inside its capsule. "The milk bacteria will grow just like the bacteria in food," Aisling explains. "The *L lactis* would be set to grow at the same growth rate as the *E coli.*"

She used log-based calculations to determine bacterial growth against time and had to develop a sensor that had the right amount of milk bacteria to achieve a colour

+ Winner of the BT Young Scientist of the year 2006, Aisling Judge, a 2nd-year student from Kinsale Community School Co. Cork, with her class mates.

change while growing at a rate to replicate the bacterial growth expected during food spoilage.

She was not allowed to use actual *E coli* in the development of her system given that it is a dangerous pathogen and requires special labs and equipment. It is also well characterised, however, and its growth rates at given temperatures are well known, so Aisling was able to fine tune her *L lactis*-based system to mirror *E coli* growth. The system could also be matched with other common food pathogens such as *Campylobacter* or *Salmonella* she said. This did not form part of her project, however, as the growth rate data were not readily available to her.

Aisling's project is typical of the wonderful research studies completed each year by the thousands of students who participate in the BT Young Scientist and Technology Exhibition. This year's was the largest yet, with just over 500 projects in the pure sciences, mathematics, technology, biology, ecology and the social and behavioural sciences.

Aisling will now go forward this autumn to participate in the EU's own Young Scientist competition, where over the years Irish students, often the youngest at the exhibition, have claimed many of the top awards.

In keeping with Ireland's long traditions in scientific endeavour, Ireland's Young Scientist exhibition is in its 42nd year. This makes it one of the longest running exhibitions of its kind in the world. Long may it continue.

+ **Kinsale Community School:** http://www.kinsalecommunityschool.ie/

List of captions and credits for illustrations

Index